Campden & Chorleywood Food
Research Association Group

Chipping Campden, Gloucestershire, GL55 6LD, UK
Tel: +44(0) 1386 842000   Fax: +44(0) 1386 842100
www.campden.co.uk

Review No. 51

STARCH HANDBOOK

Cereal and tuber starches:
their nature and performance in foods

Dr. Robin Guy

2006

Campden & Chorleywood Food Research Association Group comprises

Campden & Chorleywood Food Research Association

and its subsidiary companies
CCFRA Technology Ltd   CCFRA Group Services Ltd
Campden & Chorleywood Magyarország

© CCFRA 2006

Information emanating from this company is given after the exercise of all reasonable care and skill in its compilation, preparation and issue, but is provided without liability in its application and use.

The information contained in this publication must not be reproduced without written permission from CCFRA.

Any mention of specific products, companies or trademarks is for illustrative purposes only and does not imply endorsement by CCFRA.

Legislation changes frequently. It is essential to confirm that legislation cited in this publication and current at the time of printing, is still in force before acting upon it.

© CCFRA 2006

CONTENTS

Abbreviations		v

1.	**WHAT IS STARCH?**	1
1.1	The nature of starch polymers	3
1.2	Formation in the plant	3
1.3	The nature of the starch granule	4
1.4	The fine structure within the granules	6
1.5	The nature of crystallinity in starch	6
1.6	Macroscopic forms of starch and their location in grains and tubers	9
1.7	Non-starch constituents of starch granules	11
1.8	Starch gelatinisation phenomena	11
1.9	Starch recrystallisation	15
1.10	Starch and the glass transition	15
1.11	Commercial forms of starch used in industry	17

2.	**HOW BASIC FORMS OF STARCH ARE MADE**	19
2A	Dry milling of cereals	19
	2A.1 Overview	19
	2A.2 Wheat milling	20
	2A.3 Maize milling	21
	2A.4 Rye milling	22
	2A.5 Barley milling	22
	2A.6 Rice milling	23
	2A.7 Oats milling	24

2B	Wet milling and washing processes	25
	2B.1 Overview	25
	2B.2 Maize starch manufacture	26
	2B.3 Wheat starch and gluten manufacture	28
	2B.4 Rice starch manufacture	29
	2B.5 Potato starch manufacture	30
	2B.6 Tapioca (Cassava) starch manufacture	31
	2B.7 Minor starches (sago and arrowroot)	33

3. HOW SPECIAL TREATED FORMS OF STARCH ARE MADE — 39

3A	Heat-treated raw materials	39
	3A.1 Overview	39
	3A.2 Drying processes to reduce moisture	40
	3A.3 Complex drying for tuber products (potato)	40
	3A.4 Dry thermal processes to reduce microbiological loading	42
	3A.5 Steam treatments to denature enzymes and reduce microbiological loading	42
	3A.6 Heat moisture treatments (annealing and par boiling)	43
	3A.7 Thermal treatments of flour for cake making	43
	3A.8 Very high temperature (VHT) thermal treatments of starch for improved cooking performance	46

3B	Chemically modified materials	47
	3B.1 Overview	47
	3B.2 Degradation or conversion of starch	48
	3B.3 Chemical cross-linking	52
	3B.4 Chemical monosubstitution of starch	55
	3B.5 Pregelatinised starch (IN-starch)	56

4.	**HOW STARCH IS MEASURED**	59
4.1	Overview	59
4.2	Composition of starch	59
4.3	Granular size and shape	62
4.4	Swelling power and Solubility Indices of starch	64
4.5	Gelatinisation temperature (melting point of crystalline regions)	65
4.6	Molecular size of starch polymers	67
4.7	Starch paste viscosity	68
4.8	The rheology of starch pastes	74
4.9	Starch gel strength	75
4.10	Retrogradation of cooked starch pastes	76

5.	**PERFORMANCE OF STARCH-BASED MATERIALS IN FLUID FOODS**	79
5.1	Overview	79
5.2	Primary considerations for starch usage and selection in fluid foods	83
5.3	Selection for a production process	88
5.4	Dry mix to be cooked for immediate consumption	88
5.5	Instant starch added in cold processed sauce or dressing	90
5.6	Starch in recipes cooked in large pans to 85-90°C	91
5.7	Starch in a recipe processed by Scraped Surface Heat Exchanger technology	91
5.8	Starch in a recipe processed by canning and retorting processes	91

6.	**PERFORMANCE OF STARCH-BASED MATERIALS IN SEMI-MOIST SOFT FOODS**	93
6.1	Overview	93
6.2	Fruit pie and flan fillings	94
6.3	Cooked grains and tubers	96
6.4	Pasta	98
6.5	Noodles	102
6.6	Cakes	104
6.7	Standard UK bread products	108
6.8	Naan breads	114
6.9	Pita bread	114
6.10	Samosas	115
6.11	Spring rolls	115
6.12	Cooked meat products	116

7.	**PERFORMANCE OF STARCH-BASED MATERIALS IN HARD AND BRITTLE FOODS**	119
7.1	Overview	119
7.2	Biscuits	120
7.3	Breakfast cereals	124
7.4	Breadings	128
7.5	Breadsticks, crouton and toast	130
7.6	Snack foods	131
7.7	Products made from low moisture doughs - confectionery	137

8.	**REFERENCES**	141

9.	**INDEX**	149

ABBREVIATIONS

Full name	Abbreviation
Acetylated distarch glycerol	ADSG
Acetylated distarch phosphate	ADSP
Acetylated oxidised starch	AOS
Acetylated starch monoacetate	ASMA
Aceylated distarch adipate	ADSA
Adipic anhydride	AA
Cook-up starch	CU
Differential scanning calorimetry	DSC
Distarch glycerol	DSG
Distarch phosphate	DSP
Epichlorohydrin	ECH
based on flour weight	fw
Genetically modified	GM
Glass transition temperature	T_g
High performance size exclusion chromatography	HPSEC
High temperature	HT
High ratio	HR
Hydroxypropyl distarch glycerol	HPDSG
Hydroxypropyl distarch phosphate	HPDSP
Hydroxypropyl starch	HPS
Instant starch	IN
Low ratio	LR
Monostarch phosphate	MSP
Oxidised starch	OS
Phosphated distarch phosphate	PDSP
Phosphorus oxychloride	POC
Scanning electron microscope	SEM
Scraped surface heat exchanger	SSHE
Sodium octenyl succinate	SOS
Sodium trimetaphosphate	SMP
Temperature for melting of starch granules crystallites (also known as the gelatinisation temperature)	T_m
Temperature of the glass transition	T_g
Very high temperature treatment	VHT
w/w	weight for weight

Starch handbook

Chapter 1

WHAT IS STARCH?

Starch is a natural biopolymer found in the world's cereal crops, important tubers and many other plant sources of lesser commercial importance. It is usually present in seeds or tubers to act as a store of carbohydrate and can be used by plants during their growth phase after germination. Large amounts are used in food products from the world's largest crops: wheat, maize and rice, and lesser crops such as potato, cassava, oats, rye, barley, millet and sorghum.

Starch forms the major constituent of all these materials, representing 60-90% by weight of their dry solids. It is stored in a unique physical structure, known as the "starch granule", (Figure 1.1) that has been adapted by man to produce a wide range of foods and beverages by long processes of empirical development. Research studies have gradually unlocked the secrets of starch and made its use in food manufacture more predictable. This chapter looks at the achievements in this field related to the understanding of the nature of starch.

Figure 1.1: Light and scanning electron microscope view of wheat, maize and potato starches

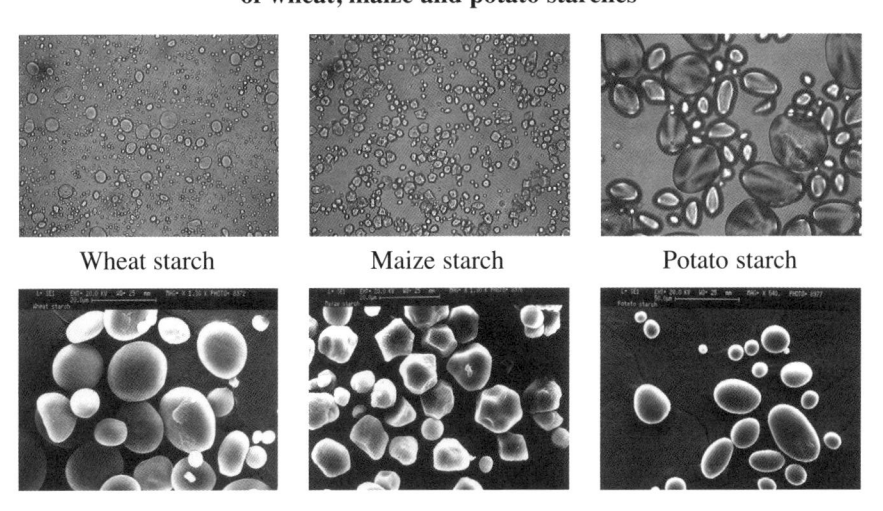

| Wheat starch | Maize starch | Potato starch |

Figure 1.2: Starch polymers (a) amylose and (b) amylopectin

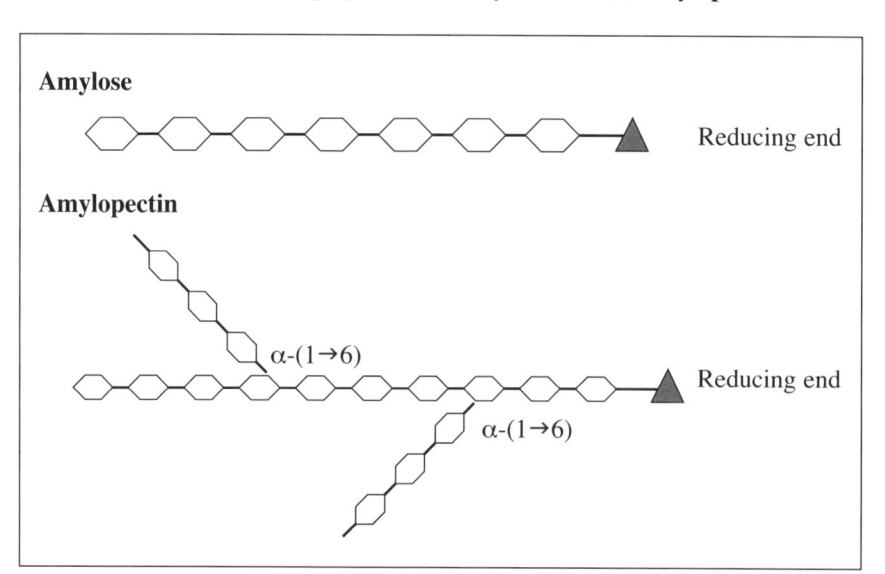

Figure 1.3: Structure of amylopectin adapted from the ideas of Hizukuri (1986)

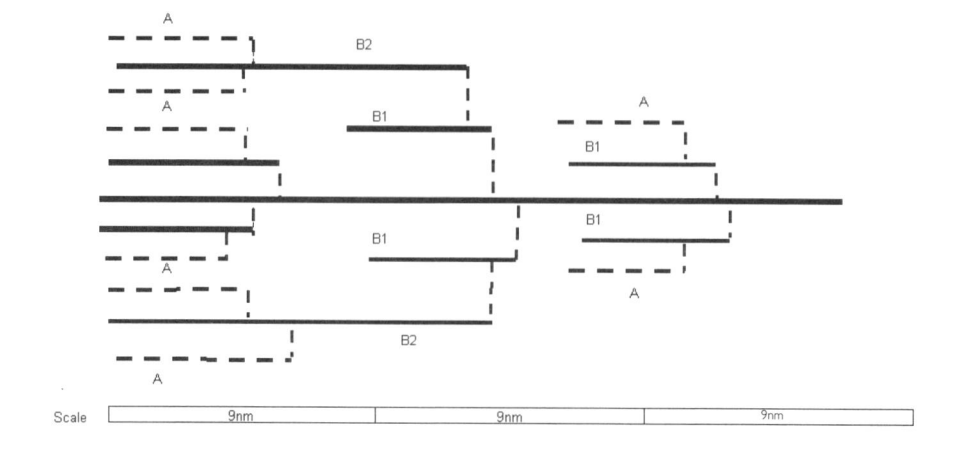

What is starch?

1.1 The nature of starch polymers

Starch biopolymers are polydisperse forms of *alpha*-glucans with linear sections linked by α-(1→4) bonds and some branch points along the chains with α-(1→6) bonds. Two separate polymer forms, known as amylose and amylopectin, are recognised in the polymer population (Figure 1.2). Both forms are also polydisperse, having size ranges of 0.1-1MDa and 1-100MDa, respectively.

Amylose is a family of long linear polymers with 99% of α-(1→4) linkages, with very few α-(1→6) branch points, and roughly 3 to 11 chains per molecule. The chains may vary in length from 200-700 glucose units.

Amylopectin is a family of much larger polymers with complex branched structures, consisting of a main chain with secondary and tertiary branches (Figure 1.3, Hizukuri, 1986; Hizukuri and Maehara, 1990). Each amylopectin molecule has about 95% of α-(1→4) linkages and 5% of α-(1→6) branch points. There is only one reducing sugar terminal for each molecule, but there are many non-reducing end groups. The side chains of amylopectin are relatively short compared with amylose, having only 18-25 glucose units. Like globular proteins, the primary structure of amylopectin (branch point distribution and side-chain lengths) appears to be responsible for the secondary and tertiary structure of the macromolecules and the relationship between structure and functional performance in starch granules.

1.2 Formation in the plant

Starch biopolymers are created by enzyme systems in plant cells from the disaccharide sucrose via glucose derivatives. Starch synthetase enzymes build up linear chains of glucose units from the non-reducing end as α-(1→4) glucan structures to form linear polymers and then add the secondary and tertiary chains at intervals to form the α-(1→6) branch points of amylopectin. These polymers are packed into complex structures, called starch granules, with sizes ranging from 1-100μm. There are several ways in which the formation of starch polymers can be varied during the polymerisation processes. The size of the polymers, their chain lengths and the amount of branching can all vary from one plant type to another and sometimes for varieties within a plant type. The proportions of amylose and amylopectin can vary over a range 15-30% amylose in wheat, to 0-70% amylose in crops such as maize, barley and rice. The 100% amylopectin forms of these starches are called waxy, because of the appearance of the cut surface of a waxy maize grain, and recent studies have also developed waxy forms of wheat starch (Graybosch *et al.* 2005).

Figure 1.4: Growth pattern of starch granules

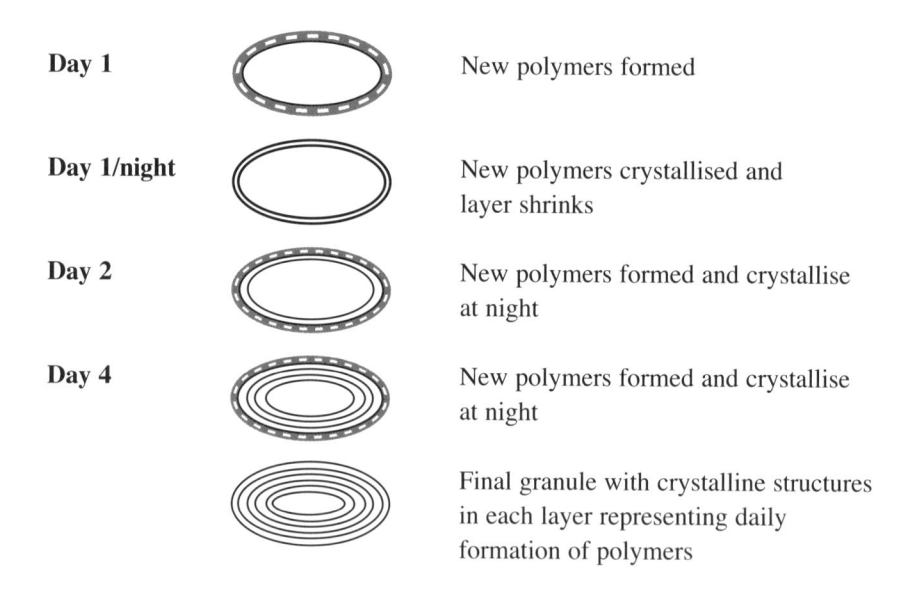

Day 1 — New polymers formed

Day 1/night — New polymers crystallised and layer shrinks

Day 2 — New polymers formed and crystallise at night

Day 4 — New polymers formed and crystallise at night

Final granule with crystalline structures in each layer representing daily formation of polymers

1.3 The nature of the starch granule

All the starch found in seeds, tubers and grains exists as starch granules. The starch biopolymers are laid down in a radial orientation in the plant cells to form granules. Biosynthesis takes place in a subcellular organelle called an amyloplast, where a granule is formed on an internal lipoprotein matrix surrounded by a lipoprotein membrane. Starch granules are formed as hollow spheres with starch polymers in the outer walls. After growing in size during the day, the walls shrink during the night and the side chains of amylopectin molecules link together to form crystalline regions. Next day, a new external layer is formed and the sequence continues the following night (Figure 1.4) The process of shrinking and growing continues throughout the deposition period and the granules grow by forming layers and eventually fill the volume within the amyloplasts. Some plants may have more than one granule growing in the same amyloplast and can form loosely packed compound starch granules, as for example in rice and oats. Others can develop small extensions to their amyloplasts in which smaller granules are produced in addition to the main large granule, as in wheat and barley, to form a bimodal population of large A and smaller B granules.

Starch granules grow from a central region, the hilium, where a cavity may remain after growth. The deposition of the layers of polymers is not always regular as some types, such as potato granules, have their hilium displaced to one end of the mature granules. Lenticular wheat starch granules develop from small spherical granules about 1-5μm diameter by first forming lips in one plane around the nucleus and then developing more normally in all directions to give the characteristic convex lens shape.

Starch deposition continues in the ripening phase of the growth of the plants, with diurnal variations. The granules grow to a maximum size in the range 5-100μm, depending on plant type. Small amounts of lipid and proteins are also found within the granules and at their surface. The amylopectin molecules in each layer link together through side chains to form crystalline regions that serve as rigid junction points throughout the granules and prevent significant swelling in cold water. Most starch granules can absorb about a third of their weight of water and remain firm and resistant to deformation because of the crystalline linkages. There is a tension in the granules due to the ring structure. As the size of the rings decreases towards the centre of the granules, the tension grows and the pressure to expand increases. If the crystalline regions are melted in excess water, the rings will push each other apart, causing the granules to swell.

1.4 The fine structure within the granules

The arrangement of the starch polymers within granules is complex, but is becoming better understood as more powerful techniques are applied. A series of annular rings were observed using light microscopy in the earliest research studies, summarised by Badenhuizen (1965). The rings were visualised by differences in density, caused by the packing of the polysaccharides. The use of techniques such as acid erosion (linterisation) to remove amorphous regions of starch polymers revealed the onion ring structures more clearly. These techniques demonstrated that the dense regions of the rings were partly composed of crystalline material. According to the latest hypotheses, these are the growth rings described in 1.3 and are comprised of zones of semi-crystalline polymers about 140-400nm thick separated by amorphous zones of the same thickness. The alternate layers of partially crystalline and amorphous regions of starch are thought to be related to diurnal growth (Donald, 2005).

The semi-crystalline zones contain about 16 clusters of amylopectin sides chains radiating from a central A-chain (Figure 1.5). Their outer branches are bound in crystalline junctions, forming crystals about 6-7nm in length, and the branch points form the amorphous areas, about 2nm in length, separating the crystalline areas. These structures are described as lamellae, alternative layers of amorphous and crystalline material. One amylopectin molecule can take part in a radial lamellae of 140nm (Donald, 2005) and will be surrounded by other amylopectin molecules to form the three dimensional ring structures. Starch granules have fairly rigid structures, but water can permeate through the amorphous areas to hydrate all the parts of the polymers except those participating in the crystalline structures.

1.5 The nature of crystallinity in starch

Since the early observations of crystallinity in wheat starch (Katz, 1930), many studies have been done to determine the nature of the crystalline structures in starch. Two distinct X-ray patterns have been observed: the A-pattern for normal cereals and the B-pattern for high amylose cereals and tuber starches. There are other starches with mixed systems of A and B patterns, sometimes called the

Figure 1.5: Crystalline structures in amylopectin showing lamellae
(based on Donald, 2005)

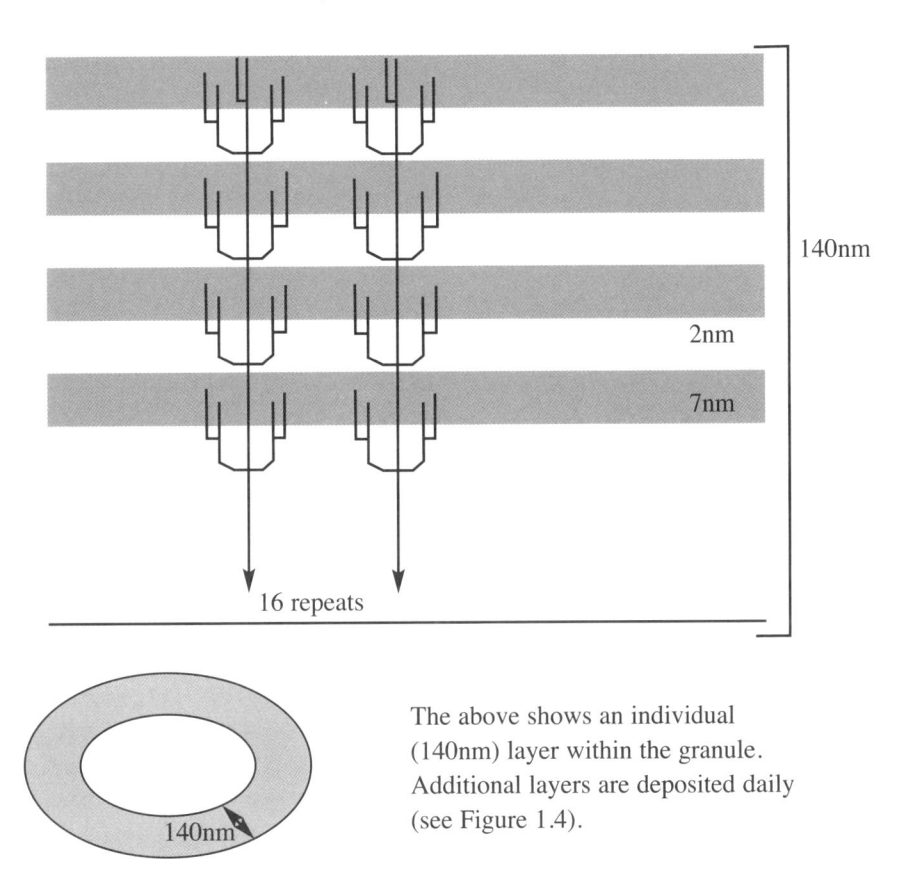

140nm

2nm

7nm

16 repeats

140nm

The above shows an individual (140nm) layer within the granule. Additional layers are deposited daily (see Figure 1.4).

C-pattern, but the basic differences between A and B types have been shown to be due to the packing of the starch chains (Wu and Sarko, 1978a; Wu and Sarko, 1978b).

In the first interactions with each other, starch polymer chains form double helices. This is common in the outer chains of amylopectin, both as inter- and intra-molecular interactions, and along the chains of amylose (Tester *et al.*, 2004). A minimum linear chain length of 10 glucose units is required to form a double helix with another section of linear starch chain. The amount of starch found with double helices is fairly high in all types of starches, varying from about 38% in wheat and maize to over 60% in rice and potato.

Six or seven double helices must be arranged in close proximity to form a crystalline junction point and create an X-ray diffraction pattern. Analysis of the X-ray diffraction patterns showed that both A and B crystal types contained closely associated double helices. However, the A-type is more densely packed than the B-type, which has 20-25% water present within its crystal, compared with 2-3% in the A type (Figure 1.6).

Figure 1.6: Arrangement of double helices in the A and B X-ray patterns of starch
(based on the interpretation of Wu and Sarko as described 1978)

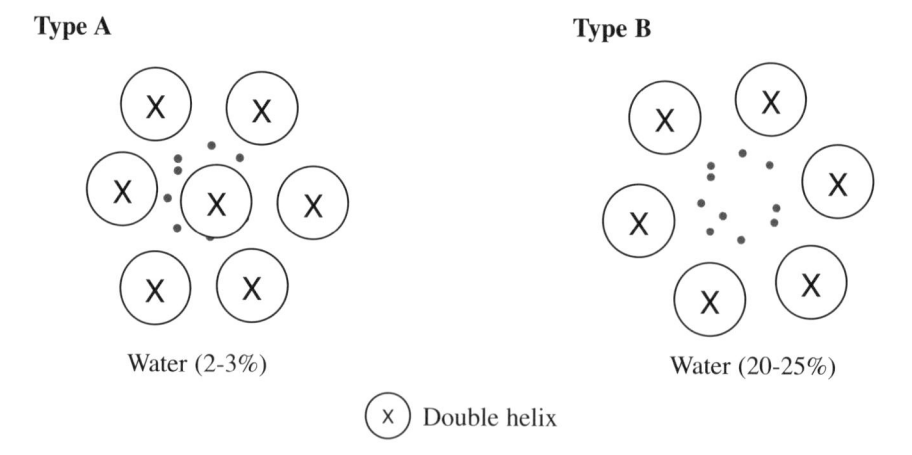

Type A

Type B

Water (2-3%)

Water (20-25%)

(X) Double helix

The amount of starch polymer found in crystallites is generally 10-15% lower than the amount present as double helices, as would be expected considering the need to align six, or seven, double helices to form a crystallite.

It is suggested that the shorter amylopectin side chains in cereal starches (14-20 glucose units) and the dry conditions in the grain (14% moisture) increase the tendency to form the densely packed A-type crystals, whereas the longer amylopectin side chains in tuber starches (16-22 glucose units) and the high moisture environment in the tuber (75% moisture) may lead to formation of the looser B-pattern, with more water molecules taking part in the crystalline structure. However, after gelatinisation in water, whereby the granules swell to twice their size and increase their moisture content to 40-50%, amylopectin in all types of starch recrystallises at ambient and low temperatures as the B-type crystallites. There are reports that at high temperatures of 45-50°C, gelatinised starch can recrystallise in the A-pattern, but these are extreme conditions for food storage.

Studies of the fine structure of wheat and rye starch by Wasserman *et al.* (2001) suggested that during biosynthesis of granules, the formation of the A-type structure is accompanied by an accumulation of crystal defects. This process leads to a decrease in the melting temperature of starches and may explain the findings of the high pressure work (Section 1.8)

1.6 Macroscopic forms of starch and their location in the grains and tubers

The size and shape of the starch granules vary from one plant source to another. For example, rice has small globular granules (1-8μm), whereas wheat has a bimodal pattern of large lenticular A-granules (20-40μm) and small globular B-granules (1-10μm) and potato has large oyster shell-shaped granules of 50-100μm, maximum dimensions. The enzyme systems that form starch biopolymers and lay them down in granules are complex. Recent studies with GM wheat have shown that the traditional shapes of wheat starch granules can be changed dramatically by manipulating the starch synthetase enzymes that affect branching in starch chains (Smith *et al.* 1997). However, commercial forms of wheat, maize, rice, potato, and cassava developed by traditional plant breeding have not changed much since they were first examined by good quality microscopes in the 1930s. Since that time, the presence of holes at the centre of granules has been reported for maize and some other cereal types. The early studies on the erosion pattern of amylases on granules suggested that potato granules were almost flawless, whereas other starches had faults, such as cracks or pores.

This information was very important for the chemists manufacturing modified starches and they pursued these features with modern scanning electron microscopes, as well as light microscopy. Observation of phosphorylated potato and sorghum starches and a hydroxypropyl analogue of waxy maize starch (Huber and BeMiller, 2001) revealed that chemical reaction patterns in starch granules were influenced by both the physical structures of starch and reagent types. In waxy maize and sorghum starches, the flow of reagents into the granule matrix occurred through channels (laterally) and cavities (from the inside outward). However, no channels were observed in potato starch granules and the reagents were only able to diffuse slowly inwards from the exterior surface through the dense starch matrix. Highly-reactive phosphoryl chloride formed most of its monostarch phosphate ester groups near granule surfaces, while the less reactive propylene oxide was shown to diffuse into the granule matrix before forming its ethers.

These studies were supported by Fannon *et al.* (1982; 2003), who observed pores in maize (corn), sorghum and millet starch granules. They showed that openings in the exterior surface led to channels that provided access to the interior of granules, and to an internal cavity. It was found that flow of reagents into the matrix of corn and sorghum starch granules occurred primarily from the channels and the cavity. Fast acting reagents formed derivatives near the surfaces of the granule, or its channels and cavity, while less reactive reagents diffused throughout the matrix of the granules before forming their derivatives. To observe granular reaction patterns within modified starch granules, starch derivatives were converted to thallium (I) salts and viewed by scanning electron microscopy compositional back-scattered electron imaging. Other techniques have been based on silver derivatives, which were reduced to metallic silver to outline the pores and viewed by confocal laser-scanning microscopy.

1.7 Non-starch constituents of starch granules

Raw materials for food preparation can be formed by separating the starch-rich materials from grains and tubers, either by milling to form flours, or by a combination of milling and washing with water to free the insoluble granules from the other materials. The three major cereals, wheat, maize and rice, have different milling processes to produce their primary raw materials: wheat flour, maize grits and polished rice. Starch granules are formed in the endosperm of the plants seeds and tubers in cells containing lipoproteins. They are often embedded in the lipoprotein membranes and surrounded by plant cell walls formed from other polysaccharides, such as arabinoxylans, *beta*-glucans and celluloses. Small amounts of proteins and lipids are found in the starch granules, both at their surfaces and in the layers of their annular rings. The proteins are mainly small albumins and globulins, and the lipids are polar types, lysolecithin and free fatty acids. These materials may have significant roles to play in the gelatinisation and swelling of starch granules. Studies of the surface proteins have suggested a role for the puroindolines in wheat hardness (Greenwell, 1994) and for the albumin and globulin proteins in the improvements of cake flour by thermal and chemical methods (Guy and Pithawala, 1981).

1.8 Starch gelatinisation phenomena

Gelatinisation under normal processing conditions

The term gelatinisation was used initially for the observation that starch granules lost their birefringence and swelled when heated in water. At concentrations of >3-5% in water, they formed pastes, or soft gels. It was noted that there was a definite temperature at which the granules lost their birefringence, called the gelatinisation temperature. At that point the X-ray diffraction pattern for starch A- or B-type crystallinity was lost and all the amylopectin in granules became amorphous. As linkages in the annular rings were lost, the granules were able to swell and even disperse in water, if the moisture content was greater than about 30%w/w. Later, it was shown that the loss of crystallinity is a water assisted melting process and that the gelatinisation temperature (T_m) for starch is affected by the amount of water present. Water is a plasticiser for the starch polymers and as the moisture is reduced below 40%, T_m increases until it is over 200°C for dry starch.

If excess water is added to native starch granules, it hydrates the polymers in the amorphous regions of the granules and causes the granules to swell a little (10%) at temperatures below T_m. At normal pressures, the transfer of kinetic energy from water molecules assists the melting process of the starch polymers in the crystalline regions. This explains the need to increase T_m and put more energy into the paste as the water content is reduced in a starch/ water system.

Figure 1.7: Swelling of starch granules

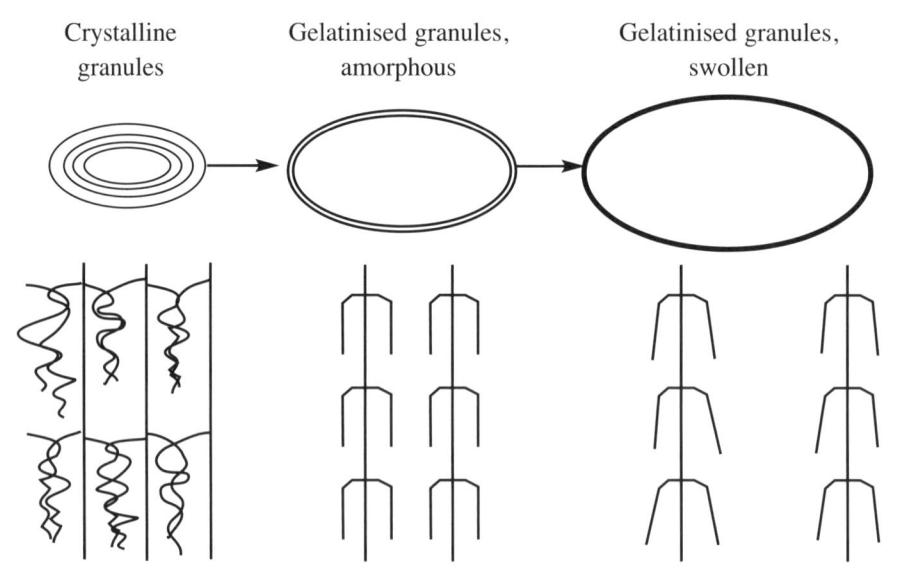

Crystalline granules → Gelatinised granules, amorphous → Gelatinised granules, swollen

Once the crystalline structures have been destroyed, water diffuses freely into the amorphous granules and they swell (Figure 1.7) as the energy in the rings is released and polymers are diluted. Starch granules heated above their T_m at <25% moisture will lose their crystallinity at 90-100°C, but will not have sufficient free water to swell. However, if an excess of water is added they will quickly swell to about twice their original size. The increase in the hydrodynamic volume of starch granules in a paste causes an increase in viscosity with a power law relationship related to starch concentration. Starch granules fill the volume of water available at fairly low concentrations (3-10%) and then form soft gels at higher concentrations. At the starch levels found in bread and cakes, the swelling is limited by the available water volume, but the granules make strong contacts with each other through the thin protein films to form a solid matrix.

Gelatinisation in the presence of high sugar levels

Starch is used in flour confectionery and sugar confectionery, where it gels in concentrated solutions of small carbohydrates, such as glycerol, glucose, fructose, sucrose, lactose and maltose. The empirical changes to the behaviour of starch in water caused by these small sugars has been observed for decades, but in more recent times a better understanding has been developed.

In the starch gelling phenomenon described above, the crystalline structures must be melted before any other changes occur. Water is the main plasticiser in starch gelatinisation in food and as its level increases the gelatinisation temperature, T_m falls. The sugars act as anti-plasticisers because as their concentrations increase, T_m also increases. Addition of a sugar to the water affects the transfer of kinetic energy to the hydrated starch chains and to the crystals. It is thought that small sugars may control a shell of the water around each of their molecules (Figure 1.8) and that this reduces the effect of the water in the systems in relation to the sugar concentration present. The effect of the small sugars increases from glycerol, through the pentoses, hexoses to the disaccharides, and is greatest for sucrose (Suggett and Clark, 1976; Suggett *et al.* 1976). The rotational relaxation of these sugars in water estimated from the dielectric constants (Table 1.1) showed slower relaxation with increasing size and number of hydroxyl groups. This suggested that the hydration shells increase in size with increasing molecular weight of the sugars up to maltotriose. Therefore, the removal of increasing amounts of water from the kinetic interaction with starch could be responsible for the increase in T_m with sugar concentration.

Figure 1.8: Hydration shells of sugars

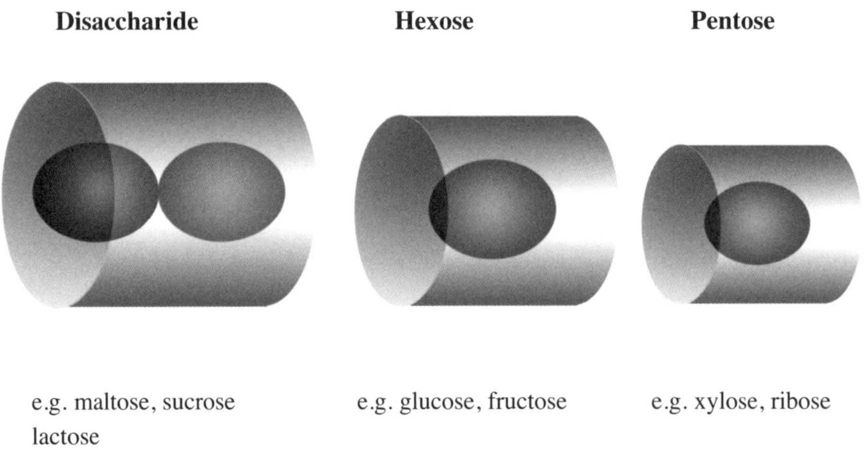

Disaccharide	Hexose	Pentose
e.g. maltose, sucrose lactose	e.g. glucose, fructose	e.g. xylose, ribose

Gelatinisation under high pressure conditions

The use of high pressure technologies, where the starch-in-water systems are subjected to up to 900MPa (9,000 bar), was shown to induce gelatinisation at ambient temperatures (Bauer *et al.*, 2005; Muhr and Blanshard, 1982). Observations of starches under pressure at high water levels (Hibi *et al.*, 1993; Gomes *et al.*, 1998) showed that there was a loss of crystallinity, measured by DSC, X-rays and birefringence in light microscopy, over a period of 30-60min. At ambient temperature, the different starch types required increasing pressure to gelatinise their granules, in the order wheat, maize and potato (Douzals *et al.* 1996). The starches with A- and C-type X-ray patterns were gelatinised at lower pressures than B-types. The gelatinisation appears to occur through hydration of the amorphous parts of the starch causing stresses on the crystalline regions, which leads to further hydration in those regions and eventually the break down of the crystallites. The A-type crystals have been reported to be less perfect and more easily disrupted than the B-type crystals. There are significant differences in high pressure gelatinisation compared to thermal gelatinisation in the extent of swelling and the damage to the polymers. These differences are reflected in the rheology of the pastes, because the pastes that are gelatinised by high pressure processing are much less viscous at the same concentration than those prepared by the optimal thermal processing.

1.9 Starch recrystallisation

Starch polymers are not stable in solution and tend to form complexes with each other and try to recrystallise. The general phenomenon for the destabilisation of soluble and hydrated starch polymers is "retrogradation". This involves the formation of double helices between starch chains of 20, or more, glucose units. Amylose is smaller and more mobile than amylopectin and can form complexes at dilute concentrations. The formation of the double helices is followed by the formation of crystalline junction points in the polymer solution, which creates a gel structure.

Amylose forms firm gels in confectionery products such as fruit gums and pastilles at high sugar and solids levels. It can also be used to create resistant starches with high levels of crystallinity to prevent hydrolysis by *alpha*-amylases.

Amylopectin forms crystalline structures in swollen starch granules and as these junction points increase in number, the swollen starch becomes a gelled structure with increasing firmness. At low concentrations, <0.5% w/w, amylopectin is more stable in water than amylose and retrogrades very slowly.

1.10 Starch and the glass transition

Starch polymers are plasticised by water and become mobile at 15-20% (w/w) moisture content, forming fluids and soft gels that are viscoelastic, and have textures varying from soft to firm. It was noted that as the water is removed from a gelatinised starch polymer system, it became firmer, passing through fluid, soft gel, rubbery and leathery states before finally becoming hard and brittle. If the final form was present in the form of a thick sheet, it would be recognised as a sheet of glassy material.

The glassy state of polymers such as starch and other polysaccharides and proteins is related to temperature, moisture and the amount of physical movement of the polymers in the food system. The glass transition temperature, T_g, is judged as the point when the texture of the starch system becomes brittle and glassy. This point can be estimated by thermo mechanical analysis (TMA) or differential scanning calorimetry (DSC). As water is removed from a product, its T_g will increase; conversely, as water is added T_g will decrease. Therefore, the value of T_g is about 5°C for bread and 50°C for a cracker biscuit with a moisture content of 3.5%.

Table 1.1: Effects of small sugars on the gelatinisation (T_m) and glass transition (T_g) of starch

Solution	T_g (°C)	T_m (at 50% concentration in water)(°C)	Dielectric constant τ (ps) (at ca. 30% concentration in water)
Water	-140	62.0	15
Glycerol	-65	64.0	-
Ribose	-47	70.4	92
Xylose	-48	74.4	99
Glucose	-43	83.8	112
Mannose	-41	80.3	100
Maltose	-30	85.0	111
Sucrose	-32	94.4	130

T_g is the temperature for the maximal freeze concentrated solute-water matrix, surrounded by ice crystals, formed by each sugar (from Levine and Slade, 1988)

The glass transition represents a change in physical form of a starch-based matrix that affects both texture and the diffusion of small molecules, such as flavours, within the starch glass. At temperatures <T_g, the diffusion rates in the glass are extremely slow and compounds are trapped within its structure until the glass melts and becomes fluid at higher temperatures. If sugars are present in the starch/water system, they act as plasticisers for the starch polymers and reduce the T_g at which the system becomes a glass. The effects of different sugars are in reverse order to their effects on starch T_m, with sucrose having the least effect and glycerol the most (Table 1.1).

1.11 Commercial forms of starch used in industry

Starch is used in the food industry as a basic nutritional component of many foods, such as bread, pasta, noodles, cooked rice and potato, where it supplies energy. Moreover, it also plays an important role in forming the structure and characteristic eating qualities of foods, such as in soups, sauces, gravies, custards, flans, desserts, baked products, pasta, noodles and coated crumb products. There are many other products that also rely on starch for their physical characteristics and quality.

Starches may be supplied in many forms from native sources in seeds, tubers and roots. In addition, special modifications have been made to improve their functionality in certain products. This chapter introduces the reader to the various forms of starch and then the text leads on to show how they are prepared in various forms, measured and used in many types of food products. In Table 1.2, a brief survey is made of the whole range available to the food industry.

In most of these raw materials, the starch is present in its granular form and as we move down the list, the granules are released from the plant cells until they are free as 'separated starches'. The finer materials such as flours and starches can be 'improved' by special treatments with heat, or chemical substitution, to give better functional performance in certain products.

The list shows a range of materials from different plant origins, which provide different starch granule sizes and compositions. The most well known difference in composition for starches is the variation in amylose content, from low levels of 1-2% in waxy maize, through 'normal' levels at 20-30% and up to high amylose starches at 50-70% amylose. These materials offer special properties and have found important niche markets in the broader field of starch supplies.

Table 1.2: Sources of starch supplied to the food industry

Basic starch-rich materials	Main products	Other types
Whole grains	Maize, rice, wheat, rye	Barley, millet, oats, Sorghum
Coarse fragments of grain >800µm	Maize hominy grits, rice cones	Oat flakes
Medium fragments of grains >350µm	Maize fine and medium grits, ground rice	
Fine fragments >150µm	Maize polenta, Durum wheat and bread wheat semolina, rice flour	
Very fine fragments <150µm	Wheat flour, maize flour	Arrowroot

Treated starch-rich materials	Main products	Other types
Precooked granules and flakes	Potato granules and flakes, flaked rice and maize	Wheat bulgur, sago pearl
Separated starches	Maize, wheat, rice, potato, cassava (tapioca), sago	Waxy maize, waxy rice, high amylose maize, waxy wheat
Dry heat treatment	Wheat flour and maize, rice and wheat starches	Tapioca starch
Moist heat treatment	Wheat flour, maize grits	Parboiled rice and its flours
Chemical modification	Maize, wheat, rice, potato, cassava (tapioca) starches	
Acid and enzyme modification	Maize and wheat starches	Other starches

Chapter 2

HOW BASIC FORMS OF STARCH ARE MADE

2A DRY MILLING OF CEREALS

2A.1 Overview

The major sources of starch used by the food industry are cereals and tubers.

- Cereals are harvested as fairly dry grains from the plants and may be processed by milling into smaller pieces and fine flours.

- Tubers are high in moisture when harvested and are processed by cutting and washing to produce starches. In the case of potato, a large amount of the wet material is also cooked during the processing to produce pregelatinised forms of the starch.

In this section only the milling of cereals is considered. This is a large subject because each cereal has its own form of milling process, developed over many years by the industry supporting that crop. The following sections briefly outline the processes used to manufacture the derivatives of wheat, rye, barley, maize, rice and oats. All cereal grains have roughly the same physical forms, with an outer coat called the husk, or hull, and an inner berry with a bran coat surrounding the starch-rich endosperm. The extraction of the valuable endosperm is the main aim of most of the milling processes, but several factors change from cereal to cereal in the shape and size of the grains and the physical nature of the separate layers and their adhesion to each other. Therefore, a number of different milling processes have been developed to maximise the end-product quality and yield from each cereal type.

2A.2 Wheat milling

The grains of modern wheats are free from husks when harvested and have an approximate composition of outer bran coat (15%), embryo/scutellum (2.5%) and endosperm (82.5%). In practice, after cleaning and conditioning the grain to moisture levels of 15-17% w/w, the endosperm is removed from the other fractions by roller milling in two main stages.

The first set of rolls are fluted and apply shear to the grains, breaking them open and scraping the endosperm from the bran, mainly as large particles called semolina. After sieving, the bran is returned to successive sets of rolls to remove more endosperm, which is again separated by sieving, to be used in the production of white flours.

The second set of rolls is used to reduce the size of the endosperm particles from semolina fractions to <150μm as flour. At the same time some further separation of particles of bran and embryo is possible to produce flour that is low in these materials. The reduction rolls are smooth and are operated with gradual reduction of the roll gaps to give smaller particles.

Flours can be made up by blending the streams of milled fractions and the recombination of separated endosperm streams in the reduction process to achieve significant effects on flour composition and quality. Flours can either be manufactured from mixed grists (blends) of different wheats, or by blending flours from different milled fractions of a single variety wheat.

White flours are blended from the separated streams of endosperm and contain <2% bran particles. For wholemeals, bran is ground to a smaller size (300-1000μm) and blended back into the separated endosperm to give a level of about 10-11% in the flour.

Wholemeal flours can also be made by reducing whole grain to a fine particle size <200μm in hammer mills using fine screens (0.8mm) and by grinding between stones in one or two stages.

2A.3 Maize milling

The family of cereals *Zea mays* is grown in most countries throughout the world. However, it requires warmer climates than are found in the temperate zones of Northern Europe to grow to maturity. The grains contain two types of endosperm, a hard vitreous outer layer and a soft mealy core. There are several important types of maize but two are widely used for food manufacture:

- Dent maize, *Zea mays indentata*, is also called "field" maize. It has kernels that contain both hard and soft endosperm that become indented at maturity.

- Flint maize, *Zea mays indurata*, has hard, horny, rounded, or short and flat kernels with the soft endosperm completely enclosed by the hard outer endosperm. In most other respects, it is similar to dent maize and is used for the same purposes.

Maize grains are formed on a cob and must be detached from the core before milling by either wet or dry processes. The maize grains have an approximate composition of bran (8.7%), embryo/scutellum (11.7%) and the two types of endosperm (79.6%). The outer endosperm is vitreous and hard with tightly packed polygonal starch granules. The inner floury endosperm has loose packing between protein and starch, with spherical starch granules. Initially in the milling process, the embryo / germ is removed to obtain the valuable by-product, maize oil.

Degermination may be carried out with a number of systems, including fluted roller, conical abrasive and impact mills. The maize grains are conditioned by adding water to raise the moisture content to 18-22% w/w, depending on the process. The best systems will give a good separation of the germ, bran and flour from the larger vitreous grits, but the other materials required might not be well separated. All the materials are dried before they are used in the second stages of the process.

The large grits are reduced in size by roller milling, using a series of break rolls to remove germ. A series of reduction rolls are used to break down the medium grits (600-800μm) to fine grits (300-600μm), polenta (150-300μm) and flour (<150μm).

Since flour is formed mainly from the soft floury endosperm, it has a slightly different composition and physical character to grits, and contains a majority of the spherical starch granules.

2A.4 Rye milling

Rye flour is similar to wheat, with an approximate composition of bran (10.0%), embryo/scutellum (3.5%) and endosperm (86.5%). It is milled by a method roughly analogous to wheat, in which cleaned grains are roller-milled to separate the endosperm from the bran. However, there are two important factors that affect the process: the endosperm of rye is weak and friable and is harder to remove from the bran. The feedstock is fed through a series of fluted break rolls where flour is separated from bran as the process continues. A series of fluted reduction rolls with gradually decreasing gaps are used to produce fine flour, fairly free from bran. Smooth rolls cannot be used because the high hemi-cellulose (mainly arabinoxylans) content of rye causes the endosperm to form flakes that would be sieved off with the bran. The yield of white flour is only 60-65% compared to 75-80% for white wheat flour. Rye flours can be made with a range of extraction rates and in a wholemeal form, as with wheat. They contain 55-60% starch in the form of large spherical and lenticular granules (25-60mm, maximum dimension) and have high levels (4-7%) of arabinoxylans.

Rye cones are coarse particles of fine or medium rye semolina that are separated by sieving and can be used as dusting flour.

2A.5 Barley milling

Barley is normally grown for the brewing industry, where the grains are malted by being forced to sprout and produce sugars for fermentation into alcohol, before drying and milling. However, there are several other applications for the endosperm of barley, which have led to the development of whole grain milling processes. Unlike wheat, most types of barley have a fibrous hull attached to the grain so that its approximate composition is hull (13.0%), bran (8.5%), embryo/scutellum (3.0%) and endosperm (76.2%). The hull must be removed from the endosperm before size reduction. This is achieved by abrasive milling with either a rotating cylinder or disc, coated with an abrasive material, operating in a static surround, such as a tube filled with barley. The grains are abraded in a series of such units, gradually losing their hull and bran and a small part of their endosperm. The final products in the main stage of this process are called "pearled" barley. This is used in soups and barley water beverages.

Pearled barley is used to make flaked barley by steaming and processing on flaking rolls. Barley flour is made by milling pearled barley with a series of rolls, firstly fluted and then smooth, as for wheat milling.

2A.6 Rice milling

Rice grains are collected from the harvest as paddy rice with their hulls attached. The kernel within the paddy rice is similar to wheat with a layer of bran (6-7%), embryo/scutellum (2-3%) and endosperm (89-91%). The aim of rice processing is to remove the hulls from the paddy rice and to produce polished white rice from the kernels. The other materials that are used in the food industry, such as rice flour and rice cones, are by-products of the main operation to produce polished rice.

The paddy rice is dehulled on shelling machines to produce brown rice. Pairs of rubber rollers are used to compress and shear the grains to remove the hulls without breaking the kernels. The brown rice is also known as shelled, cargo, husked and Loonzian rice. In the next stage it is separated from the hulls on a paddy machine using the differences in density between hull fragments and kernels.

The brown rice is debranned in conical attrition mills using iron cones surrounded by screens to remove bran fragments and produce white milled rice. This has a small amount of bran adhering to its surface and a final step is required to smooth, or polish, the surface, with a special attrition mill, fitted with leather strips to replace the abrasive metal surfaces. Rice flour is formed from the broken grains of rice, which are milled to smaller particle sizes, known, in descending order of their particle sizes, as cones, polenta and flour.

A major problem in milling rice is the lipase activity in the bran. The natural oils in the bran are exposed in the attrition mill to lipases liberated from other areas in the bran. This leads to the release by the lipase of mono- and polyunsaturated fatty acids, which are easily oxidised in air. In rice milling, it is not possible to add bran back to the milled white endosperm to form wholemeal, or brown flours, as in the wheat milling process, unless it is heat-treated to destroy the lipase. Extrusion cookers (Sayre *et al.* 1982) and other processes such as steam treatments are used for this purpose to stabilise rice bran for use in wholemeal or brown rice flours (Luh *et al.* 1991). Further modifications with proteases prior to the heat treatment have led to the development of novel functional forms of rice bran with emulsifying and moisture holding properties (Dull, 2001).

2A.7 Oat milling

Oats have an outer husk adhering to the grain, which is largely inedible and must be removed. There is a further problem in processing oats due to the lipase and high levels of lipids (6-9%) present in the grain. These components are kept separate in the grain, but mix together once the plant cell structures are broken down as the grain is milled. This leads to the formation of unsaturated fatty acids, as in rice milling, which are rapidly oxidised in air.

To overcome this problem, oats are heat-treated with steam for a short time (2-3min) before milling, raising the temperature in the grain to almost 100°C, to deactivate the lipase. The oats are dried in a kiln to reduce their moisture level to about 8% w/w, and to prepare them for removal of the husk in the "shelling" operation. The drying process can also be used to develop a mild flavour in the oats.

Dry oats are passed between a pair of circular stones, one of which is revolving. As the gap between the stones is smaller than the size of an oat grain, a mixture of husk, kernels and some broken kernels flow from the stones and are aspirated to remove the husk fragments. Shelled kernels (groats) are separated from unshelled oats over inclined table separators, known as paddy machines, and brushed to detach fine hairs, which are removed by sieving and aspiration. The polished kernels are cut transversely into four or five slices called "pinhead" meal.

Rolled oats are prepared from the pinhead meal, or even kernels, by steaming to soften the matrix of the meal before passing through flaking rolls and drying to 8-10% moisture. This process may gelatinise some of the starch and reports of up to 30% gelatinisation have been made. In recent studies the enthalpy of melting for the oat starch crystalline structures was significantly lower than for maize, rice or wheat, indicating that some gelatinisation had occurred (Guy and Sahi, 2005).

Oat flours can be prepared from rolled oats by simple milling processes such as hammer milling, or by using fluted break rolls.

2B WET MILLING AND WASHING PROCESSES

2B.1 Overview

The major sources of starch used by the food industry are cereals and tubers.

- Cereals are harvested as fairly dry grains from the plants and may be processed by milling into smaller pieces and fine flours. Pure forms of starch may be extracted from the flours.

- Tubers are high in moisture when harvested and are processed by cutting and washing processes to produce starches. In the case of potato a large amount of the wet material is cooked during the processing to produce pregelatinised forms of the starch.

Wet milling processes are used on tubers and cereal flours to produce cassava, potato, maize, rice and wheat starches (Table 2.1). The world's maize crop is 600 million tonnes per annum. Nearly 10% is made into starch or starch-derived sweeteners, making maize starch the largest starch commodity in the world. In some of the wet milling processes other important materials are collected as by-products, such as oils and protein-rich fractions. The overall economy of a wet milling process depends on the value of all the materials collected, and the cost of processing and effluents.

2B.2 Maize starch manufacture

The main forms of maize were described in 2A.3. There are also some special hybrids of Dent maize that are grown for the unusual properties of their starches. Some of these special types of maize are grown purely for their use as separated/modified starches. Examples are waxy maize, a variety whose grains have a waxy appearance when cut, and which contains >95% amylopectin. Amylomaize is another special variety, with grains having high amylose contents in the range 50-70% of the starch.

The maize grains for wet milling are cleaned twice and then fed into the steeping tanks. The maize is soaked in hot water (50°C) for up to fifty hours in a continuous counter-current process, undergoing a controlled fermentation with the addition of 0.1-0.2% sulphur dioxide. Steeping with sulphur dioxide improves the fermentation by enhancing growth of favourable micro-organisms, preferably lactobacilli, while suppressing detrimental bacteria, moulds, fungi and yeasts.

The softened kernels are broken up in attrition mills with added water to loosen the hulls and break the bonds between germ and endosperm. The oil-rich prime quality germ is separated from the ground slurry by centrifugal force in hydrocyclones and the starch-rich product stream is reground and cleaned by a second hydrocyclone separation to remove the residual germ.

After germ separation, the maize meal is finely ground in impact, or attrition mills, to release starch and gluten from the endosperm cell walls (fibres). The starch and protein particles in the system are relatively small compared with hull fragments and fibres. They are washed through screens, which retain the larger particles of hull and fibre, in a counter-current process with fresh wash water added at the last stage. The crude starch milk, containing starch, gluten, and soluble solids, is fed to a primary separator via a safety strainer and a de-gritting cyclone. The difference in density between starch and hydrated gluten makes it possible to use centrifugal force for continuous separation. Gluten passes into the overflow liquor and starch remains in the underflow liquor.

Washing with pure water refines the crude starch milk, and the fibre and solubles, including soluble proteins, are removed by counter current separation in the hydrocyclones with a minimum of water. By using multi-stage hydrocyclones all soluble materials and fine cell residues are removed in a water saving process. The refined starch milk contains only starch in pure water. This is discharged into a peeler centrifuge for de-watering and then discharged by gravity to the moist-starch hopper. From the moist-starch hopper, the starch is fed by a metering screw conveyor into a flash dryer and dried in a hot air stream. The dried starch is pneumatically transported to a starch silo ready for screening and bagging. The moisture of maize starch after drying is normally 9-12% w/w.

Maize gluten recovery

The primary separator overflow, containing mainly proteins and solubles, is concentrated on a nozzle type continuous centrifugal separator, and the overflow is used as process water. The underflow, which is mainly protein and a small amount of starch, is discharged to the gluten de-watering section. The gluten stream, containing over 60% protein, is de-watered on a vacuum belt filter, dried in a rotary dryer to approximately 10% moisture and pulverised in a hammer mill. The dried maize gluten meal is recovered at a yield of 50kg per tonne of maize, with 60% protein content. It is a valuable source of methionine and has a xanthophyll content of about 500ppm for use as a pigment in poultry feeds.

Germ purification

The germ is washed in a three-stage counter-current screen separator to remove starch, with pure water for the last stage. De-watered germ is dried in a rotary dryer to approximately 4% moisture. The remaining fibres are removed from the dried germ by a pneumatic separator and transported to the fibre silo while the germ is transported to a germ silo.

Maize oil

Mechanical presses and solvent extraction are used to extract crude oil from the germ, which is refined and filtered. During refining, free fatty acids and phospholipids are removed. The finished maize oil has a high content of polyunsaturated oils but does not develop off-flavours on cooking and frying. A typical yield of oil is 27kg per tonne of maize.

2B.3 Wheat starch and gluten manufacture

Processes developed for the manufacture of starch from wheat vary according to the demand for the two main end-products, starch and vital wheat gluten. In early processes, where starch was the main product, the gluten was often denatured to release more starch and ensure a high yield. Generally, it was found that maize was the best starting material for starch production and that wheat-based processes were less efficient.

However, in processes for manufacturing vital gluten for use in bread making and other bakery processes, only wheat flour can be used to produce vital gluten. Wheat starch produced as a by-product is sold either as starch, or converted to glucose syrup.

The first manufacturing process for vital gluten, the Martin process, used a wet gluten ball technique. In the first step wheat flour was mixed with water to form a soft dough and then the dough was washed with extra water to remove starch. The gluten was recovered in a similar manner to maize gluten with careful drying to retain vitality and the starch was recovered by centrifugation in concentrators and centrifuges. This method was first used in Australia, but has been superseded by the batter processes, which are more efficient and create less waste material in the effluent.

An example of the batter process was given by Weegels *et al*. (1988). In the first stage, dough is formed with flour and water and then diluted to give a flour to water ratio of 1:10 with vigorous mixing to form a slurry. Small gluten agglomerates form in the slurry and separation is achieved with hydrocyclones and sieves. The slurry may be pumped onto vibrating sieves, decreasing in size from one at 400µm to a series from 250 to 30µm. Gluten is retained on the 400 and 250µm sieves, bran particles on the 125 and 90µm sieves and hemicelluloses on the 50 and 31µm sieves. The starch and solubles pass through the stack and the latter are concentrated on a disc-nozzle separator.

Developments and modifications in the separation of starch and gluten continue to improve the process, improving the output (100 to 200 tonne/day), reducing effluent waste and improving efficiency. The emphasis changes depending on whether starch or gluten is the main product. In recent years, the starch has become more important in the EU and as the wheat flour process is used more for glucose syrup production, vital gluten has become the by-product. The use of wheat in the future in these processes will depend on the price of wheat and vital gluten.

The most recent process innovations use hydrocyclones or decanters to process the flour dough slurry into three crude fractions: A-starch, gluten/B-starch/pentosans and B-starch/pentosans. The gluten is refined further by washing and screening to separate B-starch and give a purer form of vital gluten. More B-starch can be recovered and added back to the total starch or kept separate as another product, the wheat B-starch fraction.

2B.4 Rice starch manufacture

Rice starch is more difficult to wash out of its endosperm than wheat or maize because its proteins are mainly insoluble glutenins and they surround the granules. The endosperm requires special treatment with alkali to release the starch granules. In a basic process for the manufacture of rice starch (Hogan, 1977; Luh, 1991), broken rice is steeped for 24h in five times its weight of 0.3% caustic soda solution at temperatures up to about 49°C. The alkaline solution is drained off and the treated rice washed with water and dried before being ground into flour. The flour is mixed with ten times its weight of 0.3% caustic soda and stirred for a further 24h. Most of the protein is removed in the supernatant fluid and starch granules are allowed to settle out. Washing with water, settling and decanting the supernatant continues for several repetitions. Finally, the washed starch is recovered by centrifugation, or filtration, free of proteins and residual alkali. The de-watered starch is dried in ovens or rotary drum-dryers and the rice cake is ground to the required size and sieved.

2B.5 Potato starch manufacture

Potatoes grow in temperate regions during spring and summer and are stored in winter. The crop is made up of small tubers weighing 100-200g, which are harvested from the soil. Loose dirt, sand and gravel are removed on a rotating screen before the potatoes are deposited in a store.

Whole potatoes are conveyed in water along a flume, passing over a trap to remove stones and grit on their way to the washing station. A counter-current process performs the washing, with fresh water added through pressure nozzles in the final step. The potatoes pass through a rotating drum where the water level is kept low, so that the potatoes do not float, and therefore they rub vigorously against each other. The rubbing is essential for the removal of fungi, rotten spots, skin and dirt from the surface.

After washing, the potatoes are shredded to release the starch from the plant cells in a machine called a "Rasper". Sodium meta-bisulphite is added to the mixture of starch, cell wall material and juice to prevent any discoloration reactions with oxygen. Starch is separated from the cell wall materials by washing in a closed system with the juice and finally discharged through conical sieves (with long perforations of 125µm). It is concentrated in a hydrocyclone to form a pumpable slurry and the crude starch milk (suspension) is then diluted and concentrated again and again using a counter current system through hydrocyclones. Incoming fresh water added in the last step and the overflow is recycled for dilution on the previous steps.

In the strong gravitational fields of a hydrocyclone and a centrifuge, starch (1.55g/ml density) settles quickly, while fibres (pulp residuals at 1.05 g/ml density) float in the wash water. The juice is directly diluted in the water and goes with the water phase.

The moist starch from the rotating vacuum filters is dried in a flash dryer with indirectly heated hot air. Potato starch is usually dried to 18-20% moisture, which is much higher than cereal starches at 9-12%. The retention of water is related to the phosphate ester groups present on potato starch.

2B.6 Tapioca (*Cassava*) starch manufacture

Cassava is a large plant with several large tubers that is cultivated in most equatorial regions and is known by many names.

Alternative names for cassava

Indonesia	Ubi kettella, Kaspe
South America	Manioca, Yucca, Mandioca, Aipim
Africa	Manioc, Cassava
India	Tapioca
Thailand	Cassava

In Europe and the USA, the names cassava and tapioca are used to describe the tubers and processed products, such as starch, respectively. Cassava tubers are very large and may contain up to 25-30% starch at the normal moisture levels of 65-70%.

Stalks are removed during harvesting and the tubers are processed in the order they are delivered to the factory, within 24 hours of harvest. In the first step, a rotating bar screen is used to remove loose dirt, sand, and gravel. The tubers then enter the washing station, where paddle washers may still be used, although rotary washers, as used in the potato industry, are more efficient. The rotary washing machines combine flushing with a low water level and continuous removal of dirt and peel. The washed tubers are conveyed on an inspection belt to the pre-cutter.

Rasping, as for potato tubers, is the first step in the starch extraction process. In order to feed the rasps properly, the large tubers are chopped into several pieces. The aim is to open all the plant cells, so that all the starch granules are released. Modern high-speed raspers use a one-pass operation to produce a slurry of pulp (cell walls), juice, and starch. After rasping, the hydrogen cyanide and cyanohydrin are released and are removed in the juice and process water. Food grade sulphur dioxide gas, or sodium meta-bisulphite solution, is added to prevent oxidative discoloration.

The starch passes through extraction sieves with the juice and the cell wall material (pulp) and is concentrated on pulp de-watering sieves that have long perforations that are only 125µm across. It leaves the de-watering sieves wet, but drip-dries to 10-15 % dry matter.

The starch slurry is diluted and concentrated in a counter current washing system, where incoming fresh water is used on the very last step and the overflow reused for dilution on the previous steps. Purified starch milk is de-watered on a continuous rotating vacuum filter, or a batch-operated peeler centrifuge. The moist de-watered starch is flash-dried with hot air to a moisture content of 12-13 % w/w.

2B.7 Minor starches (sago and arrowroot)

Sago

This starch occurs in the pith of the sago palm, *Metroxylon sagu,* which is felled after 10-15yr of growth. The tree trunks are split open to extract the pith, which contains about 50% starch, at yields of 300-600kg per tree. This sawdust-like material is mixed with water and filtered through sieves to sediment the starch in troughs. The starch is either dried for use as starch, or partially gelatinised and shaped into beads during drying to form "pearled" sago.

Arrowroot

Arrowroot starch is obtained from the root of a tropical plant *Maranta arundiinacea*. The thin roots, containing 22-28% starch, are descaled, washed and reduced in a "rasper" to a fine pulp. The starch granules are separated by screening and washed several times before drying.

Table 2.1: Properties of starches

Starch type	Granule description	Granule size, μm	Amylose, %	ΔH, J/g	T_m, °C	Critical concn
Arrowroot	spherical and polygonal	10-18 15-70	16-28	12-16	68-85	1.9
Amaranthus	polygonal	0.1-1	0.5-4	12-14	68-87	-
Chickpea	spherical and oval	8-54	32-34	-	60-75	8.3
Barley	spherical and lenticular	2-5 15-25	24-27	-	64-74	-
Cassava (tapioca)	spherical and lenticular	5-45 5-15	18-25	12-15	62-73	1.4
Maize	spherical and polygonal	5-30	25-30	11-14	62-75	4.4
Maize (waxy)	spherical and polygonal	5-30	<1	12-16	62-75	1.6
Maize (high amylose)	spherical and elongated	5-20	50-90	12-20	75-110	20.0
Oats	polygonal (compound)	3-10 (80)	23-24	3-5	50-55	-
Potato	lenticular	5-100 (mean 30)	23-26	18-21	56-66	0.1
Rice	spherical and polygonal (compound)	2-9 (7-39)	10-20 20-30	8-13	68-78	5.6
Rice (waxy)	spherical and polygonal	2-9	1-2	-	53-64	1.8

Starch type	Granule description	Granule size, μm	Amylose, %	ΔH, J/g	T_m, °C	Critical concn
Rye	spherical and lenticular	5-10 10-40	20-30	-	53-64	-
Sago	oval	20-40	28-32	-	69-70	1.0
Sorghum	round polygonal	8-20 2-6	21-30	-	68-78	-
Sweet potato	spherical and oval	3-40	15-25	12-13	64-84	2.2
Wheat A	large lenticular	15-35	20-30	7-10	52-63	5.0
Wheat B	spherical	2-10	20-30	-	55-67	-

- no value available

ΔH, enthalpy of melting of crystallites; T_m temperature of melting, or gelatinisation; Critical concentration, amount of starch required to fill the volume of 100ml of water, when swollen.

Table 2.2: Approximate composition of starch-rich materials

Cereal	Moisture %	Protein %	Lipid %	CHO %	Ash %	Crude fibre %	Food energy kcal
Amaranth	9.8	14.5	6.5	66.2	-	-	365
Arrowroot flour	11.4	0.3	0.1	88.2	0.1	-	333
Barley grain	9.4	12.5	2.3	73.5	-	-	346
Barley, pearled	10.1	9.9	2.0	78.0	-	1	350
Buckwheat grain	9.8	13.3	3.4	71.5	2.1	-	352
Buckwheat flour	11.2	12.6	3.1	70.6	2.5	-	343
Millet, raw	8.7	11.0	4.2	72.9	3.3	1	355
Maize (*Zea mays*)							
Maize grain	10.4	9.4	4.7	74.3	1.2	2.9	359
Maize flour, masa	9	9.3	3.8	76.3	1.6	1.7	358
Maize grits, dry	10	8.8	1.2	79.6	0.4	0.5	345
Maize starch	8.3	0.3	0.1	91.3	0.1	-	344
Oat groats (*Avena sativa L*)	8.2	16.9	6.9	66.3	1.7	-	378
Oats, rolled or oatmeal	8.8	16	6.3	67	1.9	1.1	372
Potato farina	10.5	10.6	0.5	78	0.4	0.2	339
Quinoa	9.3	13.1	5.8	8.9	2.9	-	363
Rice (*Oryza sativa L*)							
Rice, brown, long-grain,	10.4	7.9	2.9	77.2	1.3	1.5	347
Rice, white, long- grain,	11.6	7.1	1.9	80.0	1.65	0.6	335
Rice, white, medium-grain	12.9	6.6	1.5	79.3	1.55	0.6	329
Rice, white, short- grain	13.3	6.5	2.05.	79.2	1.65	0.5	328
Rice, white, glutinous	10.5	6.8	0.6	81.7	0.2	0.5	339

Cereal	Moisture %	Protein %	Lipid %	CHO %	Ash %	Crude fibre %	Food energy kcal
Rice bran, crude	6.1	13.4	20.9	49.7	-	10	428
Rice flour, brown	12.0	7.2	2.8	76.5	1.3	1.5	341
Rice flour, white	11.9	6	1.4	80.1	0.6	-	337
Rye (*Secale cereale*)	11.0	14.8	2.5	69.8	1.5	2	343
Rye flour, dark	11.1	14	2.7	68.7	-	3.5	338
Rye flour, medium	9.9	9.4	1.8	77.5	-	1.5	344
Rye flour, light	8.8	8.4	1.4	80.2	-	1.2	347
Sorghum	9.2	11.3	3.3	74.6	1.6	2.4	355
Tapioca, pearl, dry	11.0	0.2	0	88.7	0.1	-	333
Triticale, whole-grain flour	10.0	13.2	1.8	73.1	1.9	1.5	343
Wheat HRS	12.8	15.4	1.9	68.0	1.9	2.3	334
Wheat HW	13.1	12.6	1.5	71.2	1.6	2.3	331
Wheat HWW	9.6	11.3	1.7	75.9	1.5	-	345
Wheat SW	10.4	10.7	2	75.4	1.5	-	344
Wheat SRW	12.2	10.4	1.6	74.2	1.7	1.7	334
Wheat bran, crude	9.9	15.6	4.3	64.5	5.8	7.2	343
Wheat flour, all purpose	11.9	10.3	1	76.3	0.3	0.5	336
Wheat flour, bread	13.4	12	1.7	72.5	-	0.5	335
Wheat flour, cake	12.5	8.2	0.9	78.0	-	0.4	333
Wheat flour, tortilla	10.1	9.7	10.6	67.1	0.2	2.5	386
Wheat couscous, dry	8.6	12.8	0.6	77.4	0.6	0.6	347
Wheat bulgur, dry	9	12.3	1.3	75.9	1.5	1.8	346

- no value available

Chapter 3

HOW SPECIAL TREATED FORMS OF STARCH ARE MADE

3A HEAT-TREATED RAW MATERIALS

3A.1 Overview

The native forms of starch contain two polymer forms that contribute different properties to a cooked paste. Amylopectin hydrates well and gives good moisture holding properties, has high viscosity and recrystallises slowly, whereas amylose pastes are less viscous, but gel rapidly and give structural strength to starch pastes. These characteristics can be demonstrated by observing the differences between 100% amylopectin waxy maize, normal maize (25% amylose) and the high amylose maize (50-70%, amylose) starches.

Native starches with a mixture of amylose and amylopectin perform well in many of the major product types such as breads, pastries, pasta, cooked rice and potato. However, in the mass production of fluid food systems such as soups, sauces, desserts and cakes, native flours and starches tend to fail, as the processing becomes prolonged, or physically more severe. This failure is often due to the loss of physical integrity of the swollen granules, or hydrolytic degradation by indigenous enzymes. Therefore, special treatments have been introduced to improve starch performance. These are either based on thermal processing with steam or hot air to destroy micro-organisms and enzymes, or on chemical reactions with small amounts of active materials to cross-link or substitute hydroxyl groups on the starch chains. In recent times the thermal treatments have also been used to improve the physical performance of starch.

The first section considers all types of thermal processing.

The use of thermal methods to improve cereals and tubers covers a wide range of processing, varying from simple drying to complex heat treatments on grains and flours.

Heat treatments have been devised to treat grains, grits and flours for three purposes:

1. reducing the microbiological loading
2. reducing enzyme activity
3. improving starch granule performance

These treatments may have crossover effects in each area; for example all heat treatments will have some effect on the microbial loading and enzymic activity.

3A.2 Drying processes to reduce moisture

Starch granules are fairly robust and can withstand drying at temperatures below their melting temperature, T_m. The crystallites in the starch granules are formed from parts of large starch molecules, the remainder being amorphous polymer chains. These amorphous chains hydrate in water and become mobile so that they can effect the melting of the crystallites (lowering T_m) by assisting the melting process and absorbing energy from water as it is heated.

In high moisture systems > 50% moisture, T_m is in the range 50-75°C for most starches. As the moisture level falls and there is less energy transfer to the amorphous chains, T_m increases so that at 20% moisture, w/w, it may be as high as 115-120°C. At that point, the drying temperature may be raised to give faster drying without any gelatinisation of starch. Often a dryer can be set with a strong airflow to remove water and raise the T_m above the drying temperature before any starch is gelatinised.

3A.3 Complex drying for tuber products (potato)

Two important types of dried material, called "potato granules" and "potato flakes", are prepared from raw potato for use as water binders and dough formers in mashed potato, snackfood and meat products.

Potato granules

The term potato granules should not be confused with native potato starch granules. It refers to small pieces of dried potato of 0.5-2mm maximum dimension, compared with the native granules of 10-20μm maximum dimension.

Peeled potatoes are cut into small slices and cooked in steam cookers, or boiling water, for 30min so that their starch granules are gelatinised within the plant cells. The cooked material, at 70-75% moisture, is mixed with dry materials at 6-8% moisture from the preceding production in an "add-back" process, to give a moisture level of 35-45% w/w. During gentle blending, the average size of the pieces of potato is reduced to form the "potato granules". These are cooled to 15-20°C and held for 60min to allow some retrogradation. The moist mix is dried by blowing it over in an airlift to a fluidised bed drier, where the moisture is reduced to 6-8% w/w. Finally, the dried material is sieved and the larger particles used in the add-back.

Potato granules are easily hydrated in hot water to form the basis of mashed potato. Their starch is largely retained within the plant cell walls so that the texture of concentrated dough is created by solidly packed plant cells and is stiff, but not sticky. A small amount of monoglyceride (0.2-0.3%w/w) is added to reduce stickiness during processing.

Potato flakes

The potato flake process begins like that of potato granules with the preparation of cooked diced potato. However, the drying process is carried out by forming a slurry of cooked mashed potato with added monoglyceride (*ca* 0.3% w/w) and depositing it onto a drum dryer. A thin film of potato slurry is formed on the surface of the hot drum (100-110°C), which dries as the drum rotates and is scraped off as a thin flake of about 8-10% moisture. This is broken down by mixing and formed into a coarse powder. The starch granules are gelatinised, partly freed from the plant cells, and partly degraded so that they swell in cold water and form a slightly sticky dough. This form of precooked potato is used as a binder in several types of food products, including snacks and formed meat products.

3A.4 Dry thermal processes to reduce microbiological loading

Dry thermal treatments of grits and flours to reduce micro-organisms have been carried out using both continuous and batch equipment operating at 110-150°C, for residence times of 1-2 min for the continuous processors, and up to 20-30 min in batch processors. Moisture may be lost in these processes during heat treatment and transport to the cooling units.

Little or no gelatinisation of starch occurs during a dry thermal process unless the temperature is raised to >140°C, which can be achieved in an extrusion cooker. Some enzyme activity may also remain because their denaturation temperatures also increase at low moistures.

3A.5 Steam treatments to denature enzymes and to reduce microbiological loading

The application of steam to wheat, or flour, under pressure can raise the temperature quickly and increase the moisture a little by condensation. This increases the denaturation of enzymes and the microbial kill so that the flours may be enzyme-free with low microbial counts. Some starch may be gelatinised in local areas of high moisture, where the steam condenses. This type of process can be combined with high temperature drying to give the lowest levels of residual microbial activity.

The use of steam on wheat grains also appears to increase the T_m value for wheat flour prepared from the grain from 54°C to about 62°C.

3A.6 Heat moisture treatments (annealing and par boiling)

Annealing

If starch granules are heated in water 5-10°C below their gelatinisation temperature, T_m, their crystal structure is not melted, but is improved by annealing. The T_m measured after annealing is 5-10°C higher than before and the melting peak is reduced in breadth. The best hypothesis at present is that through the gentle energy input to the amorphous parts of the large amylopectin molecules forming the crystals, the crystal become more perfect. The alignments between the amylopectin double helices forming the crystals may be improved (Tester and Debon, 2000).

Parboiling (Luh and Mickus, 1991)

The process is used to improve the quality of polished rice and involves four stages: steeping at about 65°C for 6-7h to achieve about 30% moisture in the rice, steaming at 100-120°C for 5-10min, rapid cooling, and drying to about 12% moisture. The initial stage of the process is like annealing with improvements in crystal structures and increases in levels of starch crystallisation in the weaker parts of the grain. Vitamins flow from the bran layers into the endosperm to improve its nutritional quality, and the physical faults in the kernels are sealed so that the dried grains are less likely to shatter in the milling process.

3A.7 Thermal treatments of flour for cake making

A process for flour improvement with chlorine gas was discovered in the 1930s and used for 70 years in the USA and UK to make special cake flours for the manufacture of high-ratio cakes containing high proportions of sugar and liquid to flour. The mechanism of the improvement in the flour was traced to the starch fractions in the 1960s and doubts raised at that time about usage of chlorine led to studies of alternative processes. Only one was successful: the heat treatment of flour, patented by Russo and Doe (1968) and confirmed on US flours by

Thomasson et al. (1995). An alternative route using heat treatment on wheat, or semolina before milling as flour, was shown to give useful results for UK flours (Cauvain et al., 1979) but was not commercialised.

The early studies showed that the heat treatment of dry wheat flour at temperatures of >120°C for a short time improved its cake making performance to almost match that of a chlorinated flour. The process is at its optimum for flours that have been turbo-milled and air-classified to give a starch-rich fraction with a large proportion of free starch granules.

Since chlorination was withdrawn from use in the EU in 2001, the heat-treatment process has been scaled-up by several milling companies to supply the market. The details of the processes used have not been reported in the literature, but are based on the principle of the original findings, confirmed in more recent studies at CCFRA (Guy and Mair, 1994). They summarised their findings for the best quality heat treated cake flour as follows:

1. The flour should have a low *alpha*-amylase activity to prevent off odour and colour formation.
2. It should have a fine particle size range <75µ and a high percentage of free starch granules.
3. The flour should be dried to <8% moisture before exposing it to high temperatures >100°C to protect wheat proteins so that they hydrate and give good viscosity in cake batter.
4. The heat treatment for improvement of the starch should be >120 and <140°C for 5-10 minutes.
5. The use of an inert gas (argon) gave similar or improved performance, after treatment, than air, suggesting that the improvement is not due to oxidation.
6. The flour should be cooled quickly to prevent discoloration and off odours and flavours.
7. Moisture can be added to the flour to replace that lost during drying and heat treatment either directly on the flour, or in the cake batter.

It was also noted by Thomasson *et al.* (1995) that the cakes made from heat treated flours could be improved by increasing the cold batter viscosity with xanthan gum, or by adding L-cysteine at 200-300ppm to resolubilise wheat proteins denatured by the processing. This matches the improvement in protein dispersal caused by chlorination.

The main mechanism of the improvement has been related to the starch granules and their ability to form a strong gel structure in high sugar concentrations in a cake recipe (Guy and Pithawala, 1981). Low-ratio cakes form good structures when made with untreated flour but are rather pasty. If the flour level is reduced by 20% in the same recipe to give a high-ratio recipe, the starch has to create the same gel strength in the cake during baking to stabilise the product. Only the treated flours can give this effect due to their improved ability to form such gels. The main improvement in the recipe appears to be the ability of the starch granules to swell to a greater extent in sugar syrup after their structure had been modified by heat. Earlier studies had shown that cake flour forms stronger gels in either water or sugar syrup (Frazier *et al.* 1974) after heat treatment. Brock and Greenwell (1996) showed that acetylation of the amine groups on the surface proteins gave a large improvement similar to chlorination, or the best heat treatment. Since this is another form of chemical treatment it has not been exploited commercially.

The current hypothesis is that the starch granule's surface and any channels leading into the structure have been made more porous to the diffusion of sugar syrup. This could give extra swelling volume, more intergranular contacts and greater concentration of the interstitial egg gel (Guy and Pithawala, 1981). It was noted in a study on heat-treated flours, that increases in the damaged starch level had a negative effect on its performance in cakes (Anderson *et al.* 2000). This suggests that the volume effect of sound native granules as they swell is the critical feature of cake structure formation to create a high viscosity in the crumb.

Heat-treated flours with low protein levels (7-9%) are used in the basic cake recipes but for the products containing dried vine fruits or cherries, higher protein levels (11-12%) are used to give higher viscosities in the batter, prior to starch gelatinisation and swelling.

3A.8 Very high temperature (VHT) thermal treatments of starch for improved cooking performance

The functional effect of starch in processing has been improved in the past by chemical substitution and cross-linking, as described in Chapter 4, and also by special treatments with gases, such as chlorine, to improve their performance in high sugar concentrations.

However, a special form of thermal processing has been developed to produce similar performance to chemically cross-linked native granules (Chiu *et al.*, 1998). The new starch derivatives are made from the major starch-rich materials, as used for chemical modification: maize, waxy maize, rice, tapioca and potato.

In the new process, starches or flours are mixed in dilute alkali at pHs from 7-9.5 and then dried to very low moisture (about 1%) at <120°C, in preparation for a severe heat treatment. This ionises the hydroxyl groups and makes them reactive. Other techniques may be used but the moisture content should be reduced to the same target value. The heat treatment is carried out in a fluidised bed at temperatures of 140-160°C, preferably 160°C, for 0.5 to 3 hours. The amount of inhibition obtained, equivalent to chemical cross-linking, increases with both temperature and time.

These starches are made in two forms, the cook-up (CU) type, which has the primary heat treatment, and the instant cold water swelling (INS) type, which is processed by drum drying from an aqueous slurry as for the potato flakes, after the dry heat treatment.

The pastes formed from these starches are more stable to cooking and shear than those from native starches and can be used to provide a good viscosity with a short texture in a similar manner to chemically cross-linked starches. They are recommended for dry products, soups, sauces and gravies and for use in products that are cooked up immediately, or reheated prior to serving. The products are also reported to have some of the stability to retrogradation that can be obtained by chemical substitution with acetyl or hydroxypropyl groups on the starch polymer chains. There is little published evidence to support this claim at the moment but work is in progress to examine these starches at CCFRA.

3B CHEMICALLY MODIFIED STARCHES

3B.1 Overview

Native starch granules occur in many foods that are the great staple diets across the world: bread, pasta, rice, potatoes, cassava tubers, and other forms of foodstuffs. In some of the specialist uses developed in the food industry, such as the fluid foods described in Chapter 5, native granules are not ideal for easy manufacturing, or producing high quality products.

In addition to fluid foods that are thickened by small amounts of starch, the sugar confectionery industry uses starches in sugar syrups with high sugar solids levels to form gelled products. Normal starches are too viscous for this purpose and modifications of the starch polymers by hydrolytic degradation of the starch with chemicals or enzymes are carried out to produce suitable forms of starch.

A number of chemical pretreatments of starches have been developed to make starch granules and polymers perform better and create a better end product.

There are several groups of treatment, which can be roughly classified as:

1. degradation of the native granule and polymer structures by chemical or enzymic action,
2. strengthening the native granules by cross-linking their polymer chains to maintain granular forms during cooking, and
3. substitution of the starch chains within native granules to inhibit recrystallisation after cooking.

These processes have a long history of development and improvement over 50-70 years. In group 1, the degradation of starch began with acid thinning (linterisation) and has continued to develop with enzymes. The main groups 2 and 3 for special starches were developed in the 1960s to overcome some of the problems encountered in the use of starch as a thickener in high moisture products, such as sauces, gravies and soups. A number of processes were developed in which the starches were washed free of proteins and hemicelluloses and then allowed to react with small amount of pure chemicals and some salts in an aqueous alkaline slurry. The slurry was neutralised and washed to remove any remaining chemicals and by-products before either drying to sell as CU starch, or cooking and drying on a roller dryer, and grinding the dry paper sheet to form instant (IN) starch products.

The whole range of starch products has been scrutinised by the food additives committees in individual countries and passed through appropriate stages of animal testing to ensure their safety. In the 1970s, the UK joined the EEC and a directive was issued describing each of the permitted additives and their full method of manufacture, purity and levels of substitution. The permitted materials were given E-numbers (Table 3.1) and have been in use since that time.

3B.2 Degradation or conversion of starch

Acid thinning

Native starch granules form viscous pastes when heated in water or sugar syrups. For some confectionery products they are too viscous for use in the high solids levels found in these processes. Confectionery products are made at high solids and low moisture and require low viscosity starches that are mobile when hot and give a good set in a reasonable time when cooled. Processes were introduced to degrade the granular structures of starches to give low viscosity pastes. The oldest form of degradation was by acid treatment of washed starches at temperatures from 25-55°C for periods of 20 to 50 hours. At these temperatures only the amorphous regions of the starch granules are attacked and some glycosidic linkages are broken to weaken the granules (Jayakody and

Hoover, 2002). No significant condensation reactions occur between the small dextrins at these low temperatures, but there is an increase in the concentrations of smaller linear molecules that improves the gelling power of the starch at relatively high concentrations. The acid-thinned starches are neutralised and dried to form the low viscosity gelling starch derivatives for use in confectionery fruit gums and pastilles. Their viscosity is measured and used to determine their fluidity rating (reciprocal of viscosity). A standard fluidity scale runs from 0 for swollen starch granules to 100 for water. Generally, the acid thinned starches used in confectionery have fluidity values of 60-75 and form strong gels. The small linear molecules created from amylopectin have gelling characteristics more like amylose and can be used to set up gels in pastilles.

Dextrinisation

Alternatively, the acid thinning processes can be continued by dry roasting the acidified starch at higher temperatures: 110-130°C for white dextrins, 135-150°C for yellow dextrins and 150-180°C for British gums (Thomas and Atwell, 1998). This action completes the hydrolytic processes and causes increasing amounts of condensation reactions between small molecular fragments of degraded starch molecules as the temperature increases. The new mixture of degraded starch dextrins and larger molecules formed by condensation are called pyrodextrins. These materials are used in confectionery products, where at high solids levels they help form the textures of coatings and jelly centres.

The pyrodextrins are also widely used in glues and adhesives, where their characteristic affinity with water is ideal for sticking on labels to containers such as beer bottles. They can easily be removed with water and are biodegradable.

Bleaching and oxidation

The treatment of native starches with bleaching agents removes coloured compounds, such as carotenoid and xanophyll pigments, to produce white starch powders, suitable for tabletting or other uses where a clean white appearance is needed in a powder mix.

If the level of bleaching/oxidising reagent is increased, it will degrade the starch polymers by splitting glycosidic linkages and opening the pyranose rings, between the C_2 and C_3 positions. The reduction in viscosity and the formation of carboxyl groups increases the affinity for water. These materials can be blended with other ingredients in water to form clear viscous and tacky pastes that have useful binding properties for coatings on meat products. The hydrated carboxyl groups reduce their ability to gel and their pastes are more stable than acid thinned starches and dextrins.

Enzyme hydrolysis to form glucose syrups

The transformation of the large polymeric forms of starch into glucose, maltose and small dextrins can be achieved with enzymes such as *alpha*-amylase, *beta*-amylase, glucoamylase and possibly some debranching enzymes (pullulanase). Starch granules need to be gelatinised for efficient hydrolysis and then a combination of acid hydrolysis followed by bacterial *alpha*-amylases, or high temperature stable bacterial *alpha*-amylase alone, can be used to quickly break down the large polymers to small branched dextrins. This reduces the viscosity in the reaction vessel and produces a fluid that can be more easily processed with other enzymes.

The second type of enzymes act from the non-reducing ends of the chains to split off maltose or glucose and form maltodextrins, maltose and glucose. Using them as catalytic reactors in bound enzyme systems serves to reduce enzyme costs.

The glucose syrups are made in a range of compositions containing increasing amounts of the smaller reducing sugars as indicated by the dextrose (glucose) equivalent (DE) number. As the amount of glucose in the mixture increases with more prolonged hydrolysis of the larger dextrins, the viscosity falls and the sweetness increases for the typical products from 20, 42, 63 to 92DE. This provides a useful range of materials for food manufacture.

The application of new enzyme systems has also led to high maltose variants and to the conversion of the glucose to fructose with isomerase enzymes in a range of sweeter products. Fructose increases the sweetness of syrups and has a greater solubility in water than glucose to give increased stability.

Future prospects with enzymes

The research on enzymes that can be used as catalysts to manipulate starch structures continues to produce exciting new products.

- Linear amylose type starch chains can be built up from sucrose or glucose to a chosen size with starch synthetases, amylosucrase or phosphorylases. These materials have good film forming and gelling characteristics over a range of viscosities.
- New enzyme systems are being applied to produce the low GI forms of dextrins mentioned below, including cyclodextrin glycotransferases, which manipulate starch chains and form rings of six to seven glucose units. These cyclodextrins have found applications in binding compounds for later release in food and pharmaceutical systems.
- New products are also being manufactured from starch, in which a portion of *alpha-* (1→4) linkages are changed to *alpha-* or *beta-*(1→2) or -(1→3) and -(1→4) linkages. Only the *alpha-*(1→4) links are susceptible to digestive enzymes. These materials are manufactured from the degradation products of starch using acid or enzyme treatments to reform glycosidic linkages in the *alpha-* or *beta-*(1→2) or -(1→3) configurations. They may be used to lower the GI and energy levels in foodstuffs.
- Other amyolytic enzyme such as amylomaltase from *Thermus thermophilus* can transfer segments of *alpha-*(1→4) glucan to the ends of chains of amylopectin, to give products with much slower retrogradation characteristics.

Research on enzymes is reported on a regular basis in Europe at the "European symposia for enzymes in grain products", ESEGP.

3B.3 Chemical cross-linking

Native starch granules remain unchanged in water, batter or dough, until they are heated above the melting point (T_m) of the crystalline structures holding their granules together. This point is also called the gelatinisation temperature, because immediately afterwards water is drawn into the granules, causing them to swell and occupy more of the volume within the fluid or dough. In food systems with high water levels, such as sauces, soups and gravies, the swollen starch granules create a significant viscosity when present in sufficient concentration to make contact with each other. The paste that forms has a viscous character, but on shearing in the mouth it feels creamy and is said to have short texture. If starch is overcooked in water, so that its granules begin to deform and break down, the character of the paste will change. It becomes less viscous and more clinging in the mouth and is said to have a long or stringy texture. Increasing the starch level to redress the loss of viscosity only makes the texture worse. In more severe processes, where the cooking time is long and at very high temperature, the starch granules may break down completely and become dispersed as polymers. The resulting fluid has a very low viscosity and watery texture.

In order to stabilise the native granules during their cooking cycle, a form of chemical modification was invented by chemists called cross-linking. Native starch granules have a structure composed of a series of 'onion' rings as shown in Chapter 1. It was thought that if the starch polymers in the outer rings were linked together with covalent bonds, the structure might hold together in a more perfect three-dimensional shape and the breakdown and dispersal of polymers would be greatly reduced.

Three materials found to be successful in this function are permitted in foodstuffs:

- adipic acid anhydride (AA) to form distarch adipates
- phosphorus oxychloride (POC), or sodium trimetaphosphate (SMP), to form distarch phosphates
- and epichlorohydrin (ECH) to form distarch glycerols (DSG)

These materials are used at low temperatures with a suspension of native starch in alkali, pH 8-9 with AA, or pH 10-11 with SMP and ECH. The starch is partially ionised, but not gelatinised, and there is little swelling of the granules while the reaction takes place. A range of starches can be produced, with increasing amounts of cross-linking, to match the requirements of a range of processing conditions. The lower levels of cross-linking are sufficient for pan cooking with short boiling times, 5-10min, while the heavily cross-linked starches are required for canning and retorting processes. Even with "heavily" cross-linked starches, the number of cross-links is small, <1%, and difficult to measure. Therefore, physical methods have been used to assess the degree of cross-linking such as the Amylograph or RVA instruments. Jane *et al.* (1992) found more cross-links in the amylopectin than in the amylose, which would be predicted from the structures proposed for starch granules in Chapter 1.

It should be noted that as the number of cross-links in the starch increases, the paste viscosity of a fixed concentration of starch in water decreases. Therefore, it is unwise to use a heavily cross-linked starch unless there is a need to do so. In a severe process, with prolonged high temperature cooking, some of the cross-links break and the viscosity of a heavily cross-linked starch increases with cooking time. The choice of cross-linked starches must be made to suit the amount of cooking time, temperature and mechanical shear in the particular process in which it is being used.

Chemical cross-linking also brought some starches into general usage that were too fragile to be used in any cooking process. Low amylose starches, such as waxy maize and the new waxy potato starch, give clear pastes and gels that are excellent for pie fillings and fruit sauces, but in their native form, their granules break down very easily on cooking. After modifications with small amounts of cross-linking chemicals, waxy maize starches have became a major form of cook up (CU) and cold water soluble (IN) starches. Similar effects have been observed with potato and tapioca starches.

Table 3.1: List of E Numbers / INS Numbers for modified starch

E number	INS	Description	Uses			
	1400	Dextrins, white and yellow	soft gel	coating fluidity		adhesives
	1401	Acid treated starch	soft gel			
E1404[2]	1402[1]	Alkaline modified starch	whiteness		stabiliser	
	1403[1]	Bleached starch			stabiliser	
	1404	Oxidised starch	soft gel		emulsifier	adhesives
	1405[1]	Enzyme treated starch				
E1410[2]	1410	Monostarch phosphate			stabiliser	
	1411	Distarch glycerol			stabiliser	
E1412[2]	1412	Distarch phosphate			stabiliser	
E1413[2]	1413	Phosphated distarch phosphate			stabiliser	
E1414[2]	1414	Acetylated distarch phosphate			emulsifier	
	1420	Acetylated starch, mono starch acetate			stabiliser	
E1420[2]	1421	Acetylated starch, mono starch acetate			stabiliser	
E1422[2]	1422	Acetylated distarch adipate			stabiliser	
	1423	Acetylated distarch glycerol			stabiliser	
E1440[2]	1440	Hydroxypropyl starch			emulsifier	
E1442[2]	1442	Hydroxypropyl distarch phosphate				
	1443	Hydroxypropyl distarch glycerol				
E1450[2]	1450	Starch sodium octenyl succinate		emulsion stabiliser	stabiliser/ emulsifier	
E1451[2]	1451	Acetylated oxidised starch	soft gel		emulsifier	

Uses columns include: thickener, binder

[1] Dextrin, bleached starch and starches modified by acid, alkali and enzymes or by physical treatment are not considered as food additives in the context of EEC Directive 95/2/EC.

[2] Modified starches, Annex 1 of EEC Directive No. 95/2/EC

[3] CCFAC International Numbering System (INS) 1989

3B.4 Chemical monosubstitution of starch

The chemically cross-linked starches form smooth textured paste with good clarity at ambient temperatures, but on cooling to lower temperatures in chill storage, these pastes change slowly with time. Both the gloss and clarity of the pastes begin to fail after a few days in chilled storage as it does in those that have been through one or more freeze-thaw cycles. This is due to the phenomenon known as retrogradation, related to the reassociation and recrystallisation of the amorphous starch. Although some starch chains are covalently bound to others and cannot take part in crystal formation, the majority of starch chains are still free to recrystallise, and at the concentrations found in the swollen granules, this is inevitable.

Acetylation

For this reason, chemists developed a second form of substitution. The introduction of acetic acid anhydride or vinyl acetate into the paste of alkaline starch led to the formation of up to 2.5% of O-acetyl groups on the starch chains. These groups are stable during cooking and are present in the swollen cooked starch granules. At moderate levels of substitution, they do not alter the texture of the starch pastes, but prevent some of the starch chains from forming crystalline structures. In this way they help to retain the texture and appearance of the starch paste for a longer time during storage. They can be used in combination with chemical cross-linking to give good cooking and storage stability. However, even these pastes still change slowly with time at chill temperatures and through repeated freeze-thaw cycling.

Hydroxypropylation

The answer was to introduce a larger and more stable group than O-acetyl. Adding propylene oxide into the alkaline starch slurries created larger substituents because the ionised hydroxyl groups on the starch opened the oxide ring of propylene oxide to form hydroxylpropylstarch ethers. This form of substitution has been found to increase gelatinisation temperatures slightly, but protects the swollen starch paste from retrogradation for long periods in both chilled and freeze-thaw cycled frozen products.

Monophosphorylation

Reacting phosphorus oxychloride or sodium trimetaphosphate with alkaline starch to produce phosphated starch monoesters achieved a third type of monosubstitution. The large ionic phosphate groups can become negatively charged in food systems and stabilise the starch by repulsion between chains. This gives good viscosity and clarity in cooked pastes for acid systems such as salad dressings, tomato sauce and creams. However, the viscosity of the pastes can be affected by salts that suppress the ionisation of the phosphate groups.

Succinylation

Octenylsuccinate esters are formed by reacting alkaline starch with 1-octenylsuccinate anhydride. The monosubstitution can be used without cross-linking to decrease gelatinisation temperature, swelling power, translucency and viscosity of starch paste (Agboola *et al*. 1997). This process also changes the starch to a more hydrophobic material and develops emulsion stabilising properties for oil/water systems and has been used as a substitute for gum acacia and in stopping fat migration in cooking certain types of meat fillings.

3B.5 Pregelatinised starch

If starch granules are used to increase viscosity and thicken fluid systems such as sauces, soups or desserts, they must be gelatinised in water so that they swell to their maximum volume. This is only possible for food systems heated to a temperature $>T_m$ if native starches are used. For low temperature processing a special form of starch must be used, known as "pregelatinised" or "precooked starch". This form of starch (IN) is heated in water to melt the crystalline regions of each granule so that they swell or are ready to swell in water and then the starch paste is dried to about 10% moisture.

Three main technologies are used to manufacture such starches: hot roll cooking, extrusion cooking and spray drying.

Hot roll cooking

The starch or flour is mixed in water at 30-40% solids w/w and the slurry is fed, via one or more small diameter feeder rolls, onto a large diameter (2m) cooking roll as a thin layer (1-2mm). The cooking roll is maintained at 105-110°C with steam so that the starch slurry gelatinises and dries out to 10% solids w/w, or less, as the roll rotates. After travelling with the drum for about 50-70% of a rotation the dried "rice paper" sheet is scraped off with fixed knives and ground to a fine powder. The powder is sieved to <200µm for use in foods. This is the most common form of instant starch, although the process is limited in throughput rates.

Extrusion cooking

The starch or flour is mixed with water at about 70% solids w/w, and fed along an extrusion screw to raise its temperature to about 130°C. After extrusion, the extrudate is dried and ground to fine particles and sifted to <200µm for use in foods. An important feature of the process is to run the cooking stage at as low a shear level as possible to retain the starch granular structure. Generally, the starch granules suffer more damage in extrusion cooking than on hot rolls and both the cold and hot paste viscosities of extruded products are usually a little lower than roller dried materials. The process is used most often to manufacture pregelled flour and food products, such as baby foods, than for separated starches.

Spray drying

A spray-drying technique has been developed to give a good quality instant starch in a small hollow spherical particle. A dilute starch slurry is blown with steam from a spray dryer nozzle to become an 'instantised' dry powder.

This material is comprised of small hollow spheres and has pasting qualities that can match cook up starch performance, and is reported to be growing in usage.

Instant starches are difficult to add to water without forming lumps but recently some materials have been developed that claim to overcome this fault.

Chapter 4

HOW STARCH IS MEASURED

4.1 Overview

There are many aspects of the use of starch that may require some measurements to be made on raw materials as a quality check. These might include one or more of the following.

- composition
- granular size
- solubility and swelling power
- gelatinisation or melting point
- crystallinity
- viscosity of polymers
- paste viscosity
- gel strength
- retrogradation of pastes

This is a fairly long list, but often the starch can be characterised for a particular application with a small selection of tests.

4.2 Composition of starch

Starch may occur in a raw material as native, mill-damaged or gelatinised granules. In processed foods it will be mainly present as gelatinised granules and solubilised, dispersed and retrograded materials.

For research purposes, as in breeding new varieties, it may necessary to analyse the starch granules in great detail. They can be separated from bran, proteins and hemicelluloses by washing doughs, or macerates of the basic raw materials, and recovering the dense granules (1.55g/ml) by centrifugation. The granules can be analysed for total starch, amylose, amylopectin, lipids and proteins. Both lipids and proteins can be extracted and examined for their individual compositions.

It is harder to analyse the starch once it is processed in a food product, because it is partly solubilised and the swollen granules have almost the same density as the rest of the proteins and fibres present. There may also be interactions that bind the starch and reduce its extraction by a standard technique. Therefore, extra procedures must be used, or a simpler analysis method used to measure only total starch content and amylose levels.

4 2.1 Total starch content

The starch polymers can be solubilised by using an alkali such as sodium hydroxide, or a solvent mixture of 80% dimethylsulphoxide (DMSO) and water*. The solubilised starch is diluted to a low concentration and hydrolysed by enzymes such as bacterial *alpha*-amylase and glucamylase, or glucamylase alone, to glucose. This sugar can be measured by a standard reducing sugar method, or by glucose oxidase and a coloured dye, or by glucose 6-dehydrogenase and a nucleotide system.

Mill damaged or gelatinised starch can be measured by enzymic methods** that avoid thermal gelatinisation or the solubilisation of native starch granules with alkali or DMSO. This is not easy because the damaged or cooked starch may have retrograded and be resistant to the enzymes and, even when available for hydrolysis, may take a long time to be hydrolysed to glucose.

4.2.2 Amylose content

The most widely used method for the measurement of amylose and amylopectin ratios is based on the binding of iodine in an iodometric titration to determine

----

* Methods included in AACC methods 76-10, Swinkels 37 1-5 Staerke
**Methods included in AACC methods 76-30A and 76-31.

the iodine affinity (IA). This relies on the greater binding power of amylose and the formation of a blue chromophore for the amylose iodine complex. The starch must be treated with a solvent to remove any monacyl lipid that may complex with amylose (Morrison and Laignelet, 1983) and the titration with dilute iodine solution gives a measure of % amylose in the starch as pure amylose binds about 20% its weight of iodine and amylopectin <0.5%. The % amylose content of starch is estimated as IA/20 x 100.

4.2.3 Amylopectin

The quantitative measurement of amylopectin is not a simple analytical procedure and most results would be by difference, subtracting the amylose value from the total starch.

In a research project the starch might be examined in detail by gel permeation chromatography on molecular sieves. Amylopectin is much larger than amylose and can be separated on columns with exclusion ranges $>10^6$ and measured by a suitable mass balance or refractive index detector. The process is slow and only suited to research projects. Alternatively, the amylopectin can be de-branched with pullulanase and the short chains separated from amylose by HPSEC (Bradbury and Bello, 1993)

4.2.4 Lipids

The small traces of lipids in starch can be measured in total by acid hydrolysis and solvent extraction in ether. They are mainly polar lipids and individual types can be extracted using a solvent such as butanol/water to be separated and measured by high-pressure liquid chromatography (Morrison, 1995).

4.2.5 Proteins

The proteins of the starch granules have attracted much interest in recent years due to studies on the chlorination and heat treatment of cake flours and the finding that hardness in the wheat endosperm was related to the presence of certain proteins on the starch granule's surface (Greenwell, 1994). The measurement of these proteins is achieved by extraction from the starch using special solutions and examination by polyacrylamide gel electrophoresis (PAGE).

4.2.6 Other special features

Native starches from tubers have a small number of phosphate groups present as monoesters on the starch chains. These are easily analysed as phosphates by hydrolysis and titration or by total phosphate analysis.

4.3 Granular size and shape

The study of starch has always required techniques to examine the appearance and size of granules. Every starch source appears to produce granules that are different in shape and size range.

4.3.1 Laser diffraction and related techniques

The particle size of starches can be determined by several techniques after separation by wet milling, or washing techniques, and drying. The simplest method is probably laser diffraction or scattering spectroscopy. The particle size distribution of the starch can be measured, for example, by helium neon laser-optical diffraction spectrometry (HELOS) with equipment supplied by Sympatec Ltd, UK. The spectrometer calculates the particle size distribution of materials by evaluating the far field intensity distribution of light diffracted by a distribution of particles through a laser beam. This can be shown as a graph to display the mean size and range.

Another technique that is used for starch particle size distribution is the Coulter Counter. This is based on the sedimentation rates of starch granules in an organic liquid, which are measured automatically by the instrument.

If the starch is still attached to other endosperm components and the size range is >150μm, an air jet sieve method can be used with selected sieves or a stack of sieves to determine ranges of particle size.

4.3.2 Light microscopy

Light microscopy was used in the early studies of starches to distinguish granules from different sources (Fitt and Snyder, 1984). It can be used in a more quantitative manner to measure granule size on microscope slides with graticules of known dimensions. This technique can give approximate sizes for the globular and regular polygonal granules in maize, but is not so good for other less regular forms found in wheat, potato and tapioca. The visualisation of starch granules can be helped by the use of staining techniques, either by optical methods or by coloured dyes. Image analysis of photographs of starch samples has been used to help in developing quantitative techniques.

4.3.3 Optical staining

Optical methods depend on the variations in polymer density in the granule and the presence of crystalline regions. The onion ring structures of starch shown in Figure 1.2 can be distinguished by their density and the crystalline regions give a strong refraction of polarised light. The use of a pair of Polaroid filters (Snyder, 1984), one below the sample and one above, enables the crystalline regions to be viewed as bright areas against a black background. The characteristic pattern of the maltese cross has been observed for many types of native starch. It disappears when the starch in gelatinised and has been used as a measure of gelatinistion in extruded pellet products used to manufacture snack foods.

4.3.4 Staining with iodine and dyestuffs

An iodine/potassium iodide solution or iodine vapour can be used to show the presence of starch granules and to distinguish the low amylose waxy types from normal starches (with 15-30% amylose) and high amylose varieties. The waxy starches form weak iodine complexes with amylopectin and appear reddish brown, while the amylose in the other types form strong iodine complexes and colour the granules from blue to purple.

An aqueous alcoholic solution of 0.1% Congo red dye can be used to distinguish between native starch granules and heat gelatinised or milled damaged granules. The dye is unable to diffuse into native granules, but penetrates gelatinised and damaged starch granules as they swell in the solution. The granules become pink and opaque and are easily distinguished from the translucent native granules.

4.3.5 Scanning Electron Microscopy (SEM)

The starch granules can be examined in three dimensions by SEM techniques (Jane *et al.*, 2006), either at ambient temperature with normal SEM, or in a frozen state with low temperature SEM. In the normal systems with dry samples, the surface of the starch is coated with gold and the reflected electron beam is used to map the surface of the sample. This technique is very useful in studying free starch granules, or ones that can be extracted from food structures.

4.4 Swelling power and solubility indices of starch

One of the basic functional changes in starch during cooking is its swelling in water. This phenomenon occurs after the gelatinisation temperature has been exceeded in the cooking process. The starch granules swell to about twice their normal diameter or eight times their volume, while keeping roughly the shape of the original granules. In an excess of water they continue to swell and as polymers diffuse out become like frilly jellyfish and gradually lose all structure to become a dispersed soluble phase.

There is a very large difference in the physical effect of swollen granules and dispersed starch polymers in foodstuffs and the loss of the granule structure should be avoided. Therefore, methods have been developed to assess the state of the starch in a cooking process where the viscosity and texture are important.

The basic method is straightforward in that a preweighed sample of starch (2g) is heated in a standard weight of water (100g) at a set temperature, 85, 90 or 95°C, for a set time, 15, 30 or 45min. The suspension is centrifuged at 3000g for 30min and the supernatant decanted into a beaker and dried at 100°C for 16h. The weight of sediment is measured to assess the swelling power. The hydrated sediment is separated off and used to determine the water absorption index (WAI) as the total weight divided by the dry solids weight. The supernatant can be recovered and dried to determine the water solubility index (WSI).

The swelling process is controlled by the interactions between the starch micelles within the granules. Cereal starches have a stronger internal bonding than tuber starches and they have a two stage swelling process. The amylose plays a role in the interactions within the granules because the waxy starches swell more easily and extensively under the same cooking conditions. Tuber starches swell smoothly to a greater volume than cereal starches, and potato starch forms a stiff paste at 2% solids compared with 5-6%w/w for maize and wheat starches (see Table 1.2 in Chapter 1). The swelling process can be too extensive and the granules may break down during cooking, particularly in medium and high shear mixing. The internal binding forces can be improved by cross-linking starch polymers by forming cross-links by thermal or chemical methods. The swelling of the modified starches is limited to a predetermined volume and they can be used to give a reproducible thickening effect in a production process.

4.5 Gelatinisation temperature (melting point of crystalline regions)

The time and temperature at which the starch swelling begins is often very important to control a processing operation. A critical change in the physical form of the native starch granules occurs after the melting of the crystallites binding the starch polymers together through the amylopectin molecules. Once the crystallites are melted, the soft amorphous form of the starch polymers can swell in water. Individual granules may melt over only 1°C and then swell, but there is a much wider range of melting temperatures, 5-10°C, throughout the starch granule population.

Measurement methods used to assess the start of gelatinisation in starch are usually based on techniques that detect the loss of crystallinity. In all cases, an appropriate sample matching the recipe or test conditions being studied is heated at a constant speed while measurements are taken continuously or at short intervals to follow the loss of crystallites.

1. A light microscope equipped with a Kohfler hot stage and cross-polarised filters can be used follow the loss of birefringence in a sample of starch suspended in water. The information from a heating run can be used to determine the temperature and time when the crystallinity is lost and also observe the swelling process (Snyder, 1984).

2. X-ray crystallography has been used for many years to measure the amount of crystallinity in starch. In one system equipped with a moving film it was used to detect gelatinisation and to demonstrate the formation of the amylose lipid complexes in the region 75-80°C. However, it is a slow and difficult technique and has been largely superseded by the faster DSC techniques.

3. Differential Scanning Calorimetry (DSC) gives a much simpler and more readily quantifiable test method. It has become the standard instrument for the measurement of the melting point of the starch and is used for starch systems of all moisture levels used in the industry. The samples (20-40mg) are heated in small pans with matching heaters and platinum temperature sensors at a constant rate in an oven block. As the crystalline structures melt, the heat input required to keep the constant temperature rise must be increased and an endothermic peak is observed in the heat/flow temperature graph. Conversely, recrystallisation requires less energy and an exothermic peak is observed. Starch has a large irreversible peak for the loss of crystallinity in the amylopectin and a smaller one for the melting of the amylose/lipid complexes at a higher temperature, but no exothermic peak for rapid recrystallisation on cooling. The amount of crystallinity in starch, measured by calibrating the calorimeter with known pure standards such as

indium and zinc, is between 8 and 18J g^{-1} for native starches. An important feature of the test method used in DSC is the heating rate. This should be as slow as possible, because there is a lag time in the instrument at high heating rates that gives positive errors in the melting point determination. However, the need to process as many samples as possible in a set time may require a compromise in absolute accuracy and costs and therefore samples may be run at 5 to 10°C/min and the measured melting point may be a degree or so too high.

4. Other spectroscopic methods are available to follow changes in starch polymer water relationships that accompany the melting of the crystallites, such as near infrared and Raman spectroscopy and nuclear magnetic resonance spectroscopy. These methods offer additional information and are very important in research studies, but are not yet available in simple forms for general usage.

4.6 Molecular size of starch polymers

Understanding the performance of starch in aqueous systems may require a more precise knowledge of the molecular size of starch. This is particularly important in high shear systems such as extruders used to make molecularly dispersed starch (MDS). This is a unique physical form of starch in which all the granules have been dispersed and the molecular size of the polymers reduced to around 0.5 to 1.0MDa, according to the processing used. Viscosity techniques are very useful to assess the size of the polymers. In the methods used in the literature, starch is dissolved in a dilute alkaline solution and the viscosity measured using classical Ublehode or Ostwald capillary viscometers.

Other workers have used Gel Permeation Size Exclusion Chromatography, GPSEC, on molecular sieves to obtain more detailed information about the size of starch polymers, but this is a more complicated technique still best operated within the confines of a research laboratory (Jayakody and Hoover, 2002; van Bruijnsvoort *et al.* 2001).

4.7 Starch paste viscosity

The measurement of the physical effects of starch-based raw materials has been studied in great detail since the earliest use of instrumental methods. Paste viscosity is directly related to the usage of starch in food products and has always provided valuable insights into its performance.

Generally, the methods have involved the preparation of a starch and water paste and the measurement of its viscosity with one of the many forms of viscometer available, such as rotational machines with cylinders, paddles, parallel plates, cone and plates and various flow devices. The preparation of a starch paste and its measurement with a viscometer requires a number of operations with good control over heating and timing, which are subject to operator error and difficult to make reproducible even in a single laboratory.

Therefore, automated machines were developed, in which a paste of starch and water was cooked under a set regime and then cooled, while continuously monitoring its viscosity. Two of the earliest machines were the Brabender Amylograph and the Corn Industries Viscometer, which still give very useful information. However, they have a long operating cycle, heating the starch water mixture from 50 to 95 and cooling back to 50°C in 2-3 h and so a faster machine was required by many users.

The Rapid Visco Analyser (Calibre Instruments UK) was developed in Australia with a flexible and well controlled cooking regime. This machine has a computer controlled cooking/cooling cycle via a metal block that can be set up to suit a particular process, including recycling, and has three levels of shear input. Its standard processing regime is completed in 15min. These machines are now widely used in the industry, and cover many aspects of the use of starches such as:

- comparing the paste viscosity achieved by different starches in a particular process where the final temperature is limited, for example, to 75, 85 or 95°C
- assessing the effects of other materials, such as fat, emulsifiers and sugars, on the paste viscosity of starches.

- comparing the performance of different starches in a recipe for a sauce, gravy or soup
- testing the stability of a starch being subjected to different process regimes (such as cooking, cooling, and reheating).
- testing for the effects of enzymes present on starch viscosity

Plate 4.1: Rapid Visco Analyser (RVA) for measurement of paste viscosity

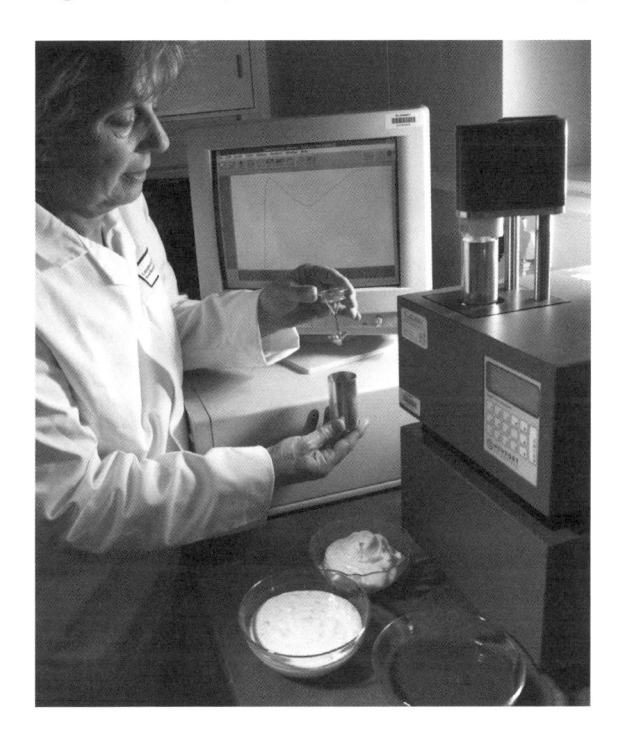

4.7.1 Example of RVA testing regime

The RVA can be used to examine the cooking of starch pastes with concentrations from 2-15%. The viscosity measurement is limited to a level of 50cps at the lower end of the range and by the gelling of starch at higher concentrations. The main contribution to the recorded viscosity comes from swollen granules. Before the granules swell, or after they are completely dispersed (highly sheared extrudates), the viscosity is very low and the graph is almost horizontal.

A typical regime for examining wheat flour is shown in Figure 4.1. For a standard comparison test a starch to water ratio of 1:10 is used. Under these conditions only the starch granules in a sample contribute to the changes in paste viscosity. The regime is run from 50 to 95 and back to 50°C over 13min. The graph rises from the baseline after 2min when the temperature is about 65°C and then reaches a peak at 5.9 min. Contacts between swollen starch granules and small amounts of soluble polymer between them determine the peak viscosity. During the holding time at 95°C, the viscosity falls because of the break down and dispersion of some swollen granules, and this effect is only reversed when the cooling cycle begins. The effect of cooling starch polymers is to increase the intermolecular hydrogen bonding and to reduce the volume fraction of the solvent so that the polymers' contacts increase with falling temperature.

Williams *et al*. (1955) described this phenomenon in terms of the glass transition temperature as shown in equation (1)

$$\text{Log } (\eta/\rho T)/(\eta_g/\rho_g T_g) = - (C_1 (T-T_g))/(C_2 + (T-T_g)) \qquad (1)$$

Where T is experimental temperature, T_g is the glass transition temperature, C_1 and C_2 are constants, ρ is density at the two temperatures and η is viscosity.

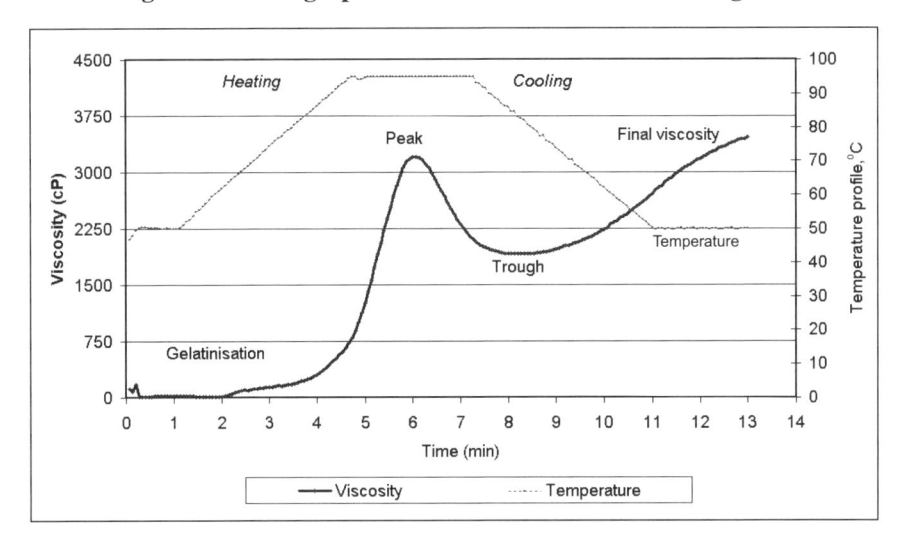

Figure 4.1: RVA graph for wheat flour in a standard regime

The final graph for the cooking test can be analysed in many ways because it is a digital trace. The standard analysis gives the values for viscosity at:

- Peak
- Trough
- Final
- Gelatinisation time
- Gelatinisation temperature

The fall from the Peak to the trough is called the Break down and the rise from the trough to the Final viscosity is the Setback.

RVA of different forms of raw material

The examination of a starch derivative in a recipe is also possible with the RVA for fluid products. The derivative should be examined in the form that is being used in the product, because both the nature and size of the particle in which it is supplied are important. In Figure 4.2 derivatives of maize milled from the same base material are examined at 10% starch equivalents. The larger particles do not swell as quickly or to the same volume.

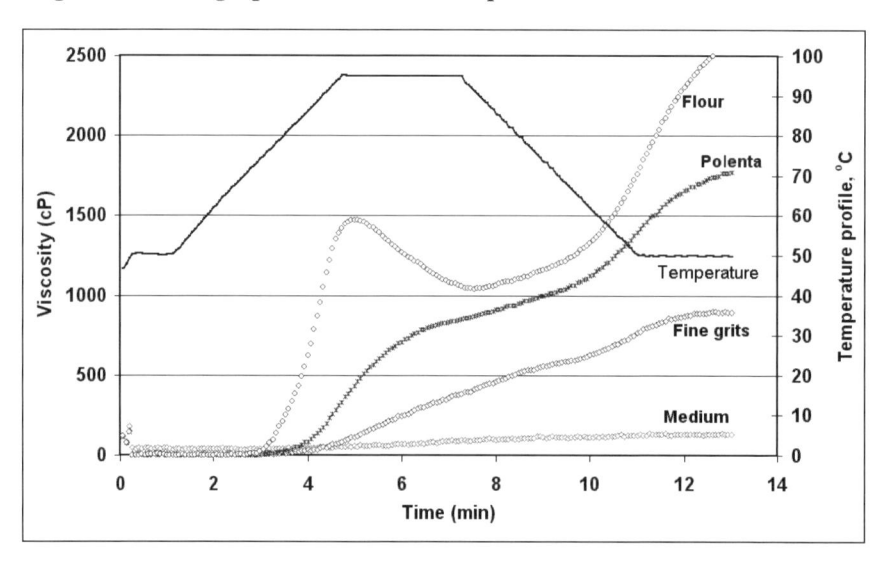

Figure 4.2: RVA graph to show effects of particle size for maize derivatives

RVA of maize derivatives in white sauce mix

The swelling of starch may be affected by the presence of other materials such as fats and salts. In Figure 4.3 the same maize flour is cooked in a recipe for white sauce at three levels of addition. At 2.5% the viscosity is very low but the viscosity is related to starch concentration as

$$\eta = AS^{3.4}$$

where S is starch and A is a constant of 0.3.

The paste does not show the normal rise as occurs in pure water, indicating a lower level of interaction between the starch polymers. It is also clear that the starch does not break down as much in the white sauce recipe as in pure water. This is probably due to the presence of lipids in the recipe.

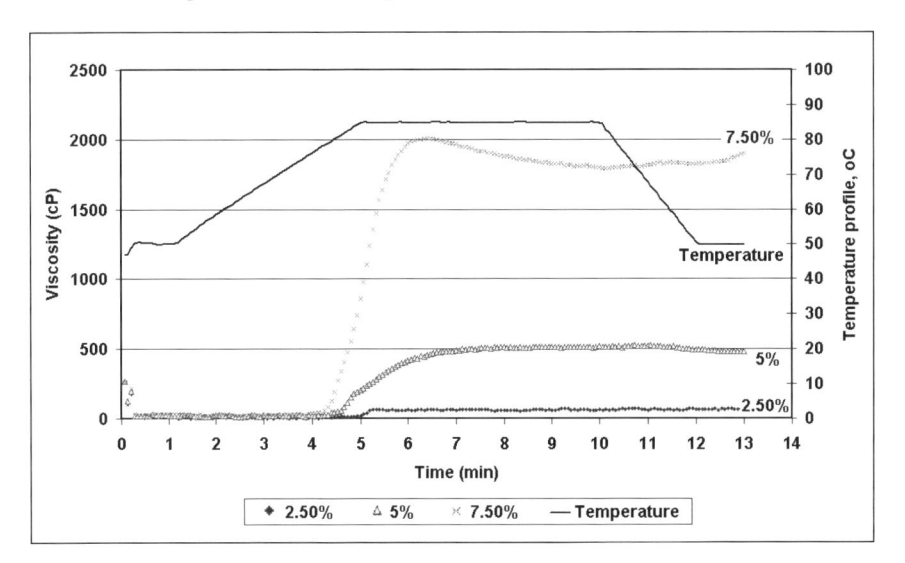

Figure 4.3: RVA graph of maize flour in white sauce

RVA of maize flour in tomato sauce

If the sauce type is changed to tomato, there is less fat but more fibre, a lower pH (4.5) and a greater viscosity at any given starch concentration compared with the white sauce.

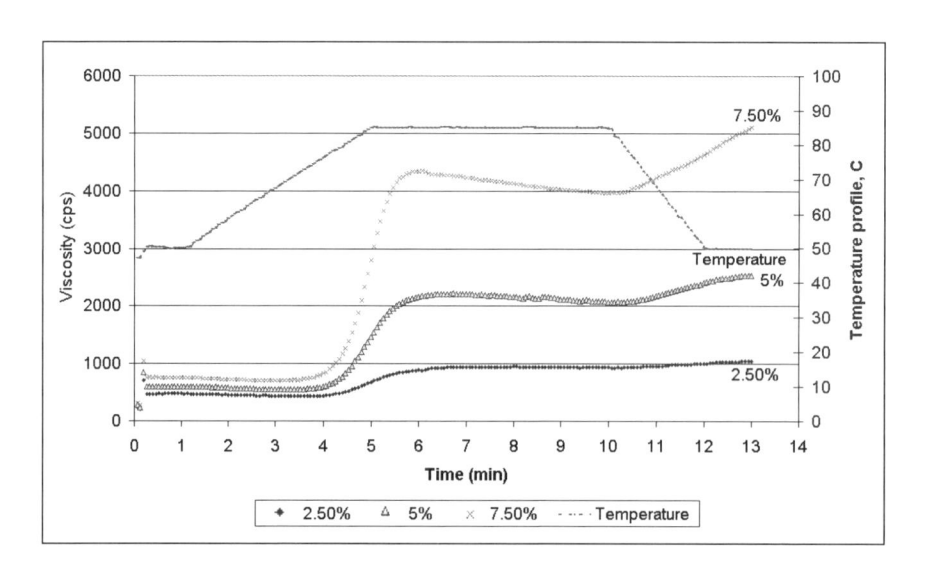

Figure 4.4: RVA graph of maize starch in tomato sauce

The base viscosity is higher for the tomato sauce and the increase is slightly slower with respect to starch concentration

$$\eta = AS^{2.7}$$

where S is starch concentration and A is a constant of 1.78.

The graph for the tomato sauce shows an increase in viscosity on cooling as obtained with the starch paste in water.

Paste viscosity measurements are very useful and can be supported by other techniques such as microscopy and more sophisticated rheological measurements.

4.8 The rheology of starch pastes

The detailed examination of the rheology of starch pastes can be carried out with a rheometer. The low deformation oscillatory rheometer, such as TA

Instruments Rheometrics ARES operating at constant strain, can give detailed analysis of the elastic and viscous moduli of a starch paste. The magnitude of the moduli and their tangent can help to measure the character of cooked or baked starch products. They will determine if it has sticky or soft or firm rubbery elastic texture and whether this characteristic changes with storage and temperature.

Plate 4.2: TA Instruments Rheometrics ARES used to assess rheology of sauces

4.9 Starch gel strength

Firm gels form on cooling in food systems where the starch concentration is sufficient for strong intergranular contacts between the gelatinised and swollen granules. The weakest forms of starch-based gels are lemon curds and custards, stronger gels are found in cakes and breads. Instruments that are designed either to compress or penetrate the surface of the gels can be used to assess gel strengths. If a small probe of well-defined geometry is driven at constant speed into a gel, the resistance on the probe can be recorded on a digital trace. The analysis of the trace can be used to determine an empirical gel strength value, or a more scientific measure such as a modulus of elasticity, depending on the test regime. There have been many studies using compressimetry to follow the firming of bread and cake on storage, and more recently these techniques have been extended to brittle foods such as biscuits and extrudates (Bhatti and Hall, 2002).

4.10 Retrogradation of cooked starch pastes

Starch pastes and gels formed after cooking or baking are not stable and will change gradually during storage. This phenomenon, called retrogradation, is related to the growth of crystallinity in the hydrated starch polymer mass (Abdd Karim *et al*. 2000).This causes changes in the physical nature of the food structure formed by the swollen starch and the control of water within these structures.

High moisture foods

In high moisture systems where the starch is present at <10% w/w and is highly swollen and partly dispersed during the cooking process, the starch may form double helices but crystallise slowly. Amylose was shown to retrograde faster than amylopectin under these conditions (Miles *et al*. 1984; Ring *et al*. 1987). The aggregation of starch polymers tends to reduce some of their control over the water and make the gel more opaque. The resulting effects are loss of surface gloss and translucency, and the leakage of water from the gel. Measurement of the separation of water from the gel after centrifugation at low g-force can be used to follow the retrogradation in systems where starch is not greater than 10% w/w (Guy and Sahi, 2006).

Intermediate moisture foods

In intermediate moisture foods such as bread or cakes (40-50%w/w) the starch polymers are mainly retained in the swollen granules. They crystallise fairly quickly, causing significant changes in the firmness of the granules and the crumb cell walls of both product types. The retrogradation is not accompanied by any visible moisture release, because the starch granules were gelatinised in a limited moisture supply and are still trying to absorb more water from the gluten layer. The growth of crystallinity in these products can be followed by DSC measurements.

Low moisture foods

In low moisture systems, such as flatbreads and soft tortillas, the starch is partially gelatinised and that part can recrystallise on existing nuclei to increase the firmness of the crumb. DSC is again very useful for following changes in the starch crystallinity.

The methods used to examine the stability of starch in different types of product should be selected according to its moisture level, as in the categories of low, intermediate and high moisture foods.

Chapter 5

PERFORMANCE OF STARCH-BASED MATERIALS IN FLUID FOODS

5.1 Overview

There are many applications where starch is used at concentrations of 2 to 10% w/w to thicken and create a paste in an aqueous fluid. The resulting products at the lower end of the scale, 2 to 5 % w/w, are usually viscous fluids that have the ability to flow under small shearing forces. Typical examples of these products are soups, pourable sauces and salad dressings sold in glass jars or squeezeable plastic containers. At higher starch concentrations they become soft textured gels, which can be spooned or cut, such as custards, cheese cakes, desserts, potted meats, fruit pie fillings and sweet and savoury flan fillings. This section concentrates on the fluid foods and leaves the last three product types to Section 6 on soft solids.

The paste viscosity developed by starch comes from its native granules, present in flours and grits, or from separated and modified starches. Native starch-rich raw materials have little viscosity in cold water at 2-10% w/w, until the starch is gelatinised when they produce high viscosities on heating the water (Figure 5.1). They are generally known as cook-up (CU) starches.

However, starch-rich raw materials that have been gelatinised, by previously cooking in water, and then dried to form a powder, are also available for use in food. These materials, which can be added to cold water to give a high paste viscosity, are called "instant" (IN) starches.

These forms of starch are used to give the main viscous characteristics for a wide range of fluid foods. Both types can be obtained in many forms, as flours, separated starches and physically and chemically modified forms of these materials, as described in Sections 2 and 3. Some examples of major products from the savoury and sweet sectors are shown below.

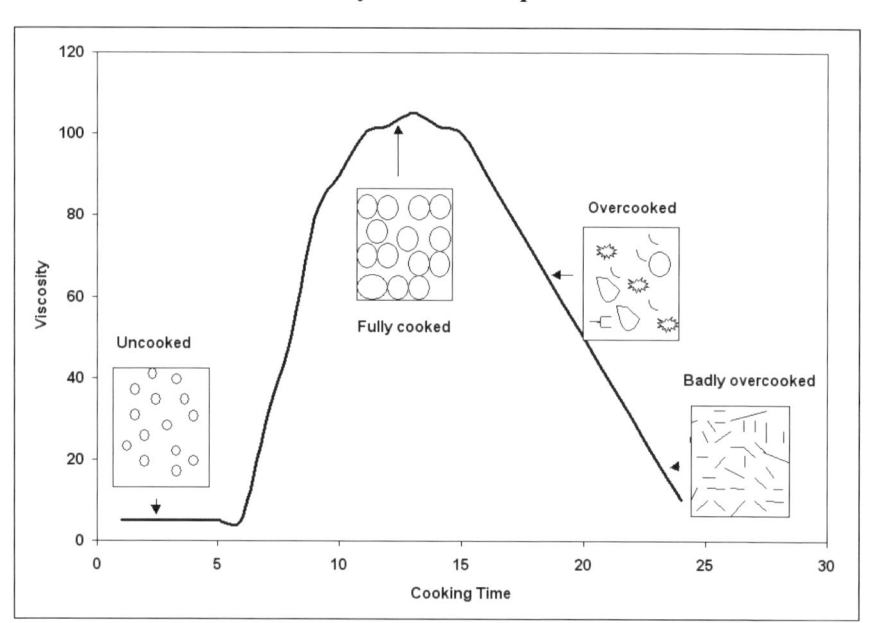

Figure 5.1: Diagram of the paste viscosity of starch granules as they swell and disperse

Savoury sauces

Savoury sauces have recipes that are based on starch rich-raw materials, but often they contain more than one type. For example, there may be a combination of flours and starches from wheat, potato, corn or rice in products manufactured by low shear processing. Further combinations may be found with modified starches replacing some or all of the native starches. These will be used as processing becomes more severe with respect to shear and/or temperature (Table 5.1). Savoury products may also contain oils and fats, dairy proteins, fibrous plant materials or salts and may range in pH from about 4.5 to 6.5. These factors may influence the swelling pattern of the starches and the final paste viscosity.

Table 5.1: Savoury products relying on starch for their consistency or texture

Savoury	Preparation conditions	Type of starch	Storage
Quick cook soup	Boiling water 1-2 min	Flours	none
Cook up soups	Boiling water 5-15min	Flours/CU modified	none
Quick sauces	Boiling water 1-2 min	Flours/CU modified	none
Cook up sauces	Boiling water 5-15min	Flours/CU modified	none
Gravy powder/granules	Boiling water 1-3min	Flours/CU modified	none
Cooking sauces	Hot water >85°C 10-20 min	Flours/CU modified	ambient long
Red and brown sauces	Boiling water	Flours/CU modified	ambient long
Prawn cocktail sauce	Cool water	Starch (IN)	ambient /chill
Ready meal sauces	Hot water >85°C 10-20 min	CU Modified/flours	short at chill or long as frozen
UHT sauces	Hot water 140°C 3-4 min	CU Modified	ambient long
Canned soups	Hot water 110-115°C for 15-20min	CU Modified/ flours	long
Meat paste/pate	Hot slurry at 80-90°C for 10-20min.	Flours	ambient long

Sweet sauces and desserts

In the sweet sector, the main differences are caused by the addition of sugar to the recipes at levels from about 10 to 70% of the weight of the water present. The high sugar levels have significant effects on gelatinisation temperatures of CU starches (Section 1.8), elevating T_m and thereby delaying the swelling process. This may be countered by using IN starches, which will swell in the sugar syrups to give a good viscosity in the cold. The textures of some of the dessert products may have specific characteristics related to lightness or heaviness in the starch paste, which can be achieved by selecting particular types of starch. Other ingredients that affect starch viscosity include the oils and fats, proteins and acids.

Table 5.2: Sweet dessert products that rely of starch for their consistency or texture

Sweet products	Preparation conditions	Type of starch	Storage
Instant desserts and cream fillings	Cold milk	IN modified	none
Cold fruit pie filling	Blended with sugar syrup	IN modified	ambient/chill
Cook up custards and desserts	Boiling water 5min	Flours/ CU modified	none
Fruit, caramel and chocolate sauces	Boiling pan to 105°C	Modified	ambient long
Yoghurt fruits	Boiling sugar syrup 5-15min	CU modified	cool medium
Fruit pie fillings	Boiling sugar syrup 1-3min	CU modified	ambient medium
UHT custards and sauces	Hot water 140°C, 3-4min	CU modified	ambient long

The aims of a food manufacturing process using starch as a thickener and texture-forming ingredient is to create a characteristic texture to suit the preference of consumers for that particular product and to maintain those qualities during its shelf-life. This is where the selection of a particular form of starch and the amount to be used in the recipe begins. It was mentioned earlier that a small number of processes do not employ any cooking but need to use an instant IN starch. The same judgements must be made about the choice of this type of starch concerning the desirable texture and the shelf-life qualities of the finished product.

Food products may be made for immediate consumption, or require reheating after being stored at ambient, chill or deep freeze temperatures. This demands remarkable flexibility in starch performance and because the more sophisticated starches are also more expensive, it is necessary to tailor the starch selection to both cost and functionality. Consequently a range of different types of starch products that are suited to different processing and storage requirements are used in the food industry.

The final equation for selection is related to

$$\text{Selection score} = (\text{Cost} \times \text{concentration}) \times (\text{Performance})$$

5.2 Primary considerations for starch usage and selection in fluid foods

Good selection and observation of some basic rules for their usage enhance the performance of starch-based materials in food systems. Before discussing the details of selection for different types of processes, this section provides some advice on using all types of starch ingredients in food manufacture. There are three main factors that need to be understood and acted upon to make the best use of a starch in creating and maintaining a desirable eating quality throughout a product's shelf-life.

1. The addition of starch to water
2. Using the cooking process to create optimum paste texture
3. The storage conditions and time.

5.2.1 Addition of starch derivatives to a high-moisture recipe

In order to produce a good texture in a product, a chosen starch derivative must be dispersed and suspended in water prior to, or during, the cooking process at a temperature 5-10°C below its gelatinisation temperature. This allows the individual powder particles to disperse prior to gelatinisation and prevents the formation of gelatinous clumps or aggregates of cooked starch. Once the starch has gelled and clumped, it is hard or virtually impossible to disperse the lumps. Badly cooked starch gives much less viscosity than expected.

- This is particularly important with pregelatinised starches, which are difficult to disperse even in cold water. They must either be dispersed in other ingredients, or added to cold water in the vortex of a high-speed stirrer or static mixer unit.

- Native wheat flour has similar problems because its proteins tend to form a sticky mass of gluten on first contact with water. This problem can be avoided by mixing a concentrated paste in cold water and then diluting this out to the required level, or by adding flour to water in a mixer with a powerful shearing head to create a vortex, as with instant starch.

- Other materials such as maize and rice flours and separated starches can be easily dispersed in cold water and poured into an aqueous mix at temperatures below the gelatinisation temperature. However, in the uncooked state they are much denser than water and need to be kept suspended by stirring until they gelatinise and swell in water, thereby lowering their density. Otherwise they may sink and form a highly viscous paste at the base of the cooking vessel and a dilute low viscosity starch paste at the top of the cooking vessel.

Some innovations claim to have treated instant starches so that the are easy to mix into water at any temperature. The details of the invention are not clear yet but they may be similar to the products made by coating hydrocolloid gums with emulsifiers.

5.2.2 Cooking to give a characteristic texture

The cooking of any CU starch derivative begins when it is suspended in water and reaches its gelatinisation temperature. Up to that point, the starch granules in CU starch-rich materials are hard and strong and will withstand very high shearing forces in a mixer.

Gelatinisation temperatures may vary with starch type by as much as 10-20°C, and increases in particle size and the effects of other ingredients may raise them still higher (Pritchard, 1992; Guy and Sahi, 2005). Immediately after their gelatinisation point (T_m) is exceeded, the amorphous granules swell and paste viscosity increases very quickly in the fluid. The starch granules swell to make contact with each other (Figure 5.1) and at the levels used in sauces and gravies (4-5% w/w) they form a fluid paste with viscosities of 20-50 poise.

The amount of starch required to fill the volume of water available is known as the ***critical concentration*** and varies for each type of starch (see Section 2, Table 2.1 and Figure 5.2). At the point where the starch granules swell to fill the volume of the water, the viscosity begins to rise fairly quickly due to the intergranular contacts and resistance to flow. Potato starch has the greatest ability to fill space in a liquid with a critical value of 0.2, followed by waxy maize and rice at 1.6, and wheat at 5.0g/100g. At the ***critical concentration*** the starch forms a viscous paste, but the texture varies a little in relation to the amount of solids in the paste. Potato starch and the waxy starches have lighter textures than wheat or normal maize starches.

Figure 5.2

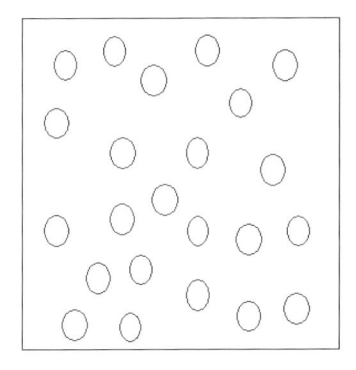

Native starch granules in a standard volume of water

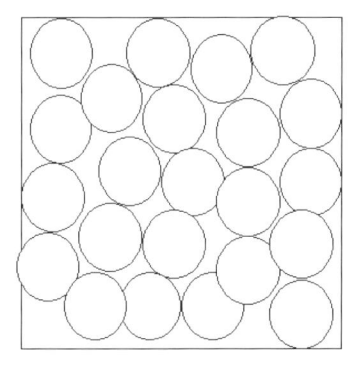

Swollen gelatinised starch granules in the same volume of water

Other ingredients in the recipe may contribute to paste viscosity, so the level of starch must be adjusted for the recipe being used. At this point the paste will be smooth and creamy with a short texture, i.e. it divides easily as a spoon is drawn through it, and does not cling in an elastic mass to the spoon. Particular care must be paid to any starch-based material that has *alpha*-amylases present because this enzyme will hydrolyse starch molecules and reduce the paste viscosity provided by swollen granules significantly during cooking. The separated starches and modified starches will have little, or no, enzyme present, but natural flours, dried egg and milk powders and other natural ingredients may have varying levels present from indigenous sources, such as micro-organisms, unless specially heat treated to destroy their activity (Section 3A).

- Unmodified starch derivatives such as potato starch, maize (corn) flour and wheat flour can be used to give a good texture suited to savoury sauces and gravies. Potato starch has the highest swelling power, but its pastes are also fairly fragile. All native starches, when cooked on a large scale, can easily suffer from shear damage and over-cooking. The starch granules continue to swell and weaken in excess water and may break down when the cooking times are exceeded, or if the cooked pastes are held at 70-75°C in a safe area before usage in the final products. Subsequently, the product loses viscosity and their texture changes from short creamy fluids to thinner slightly clinging elastic pastes. This change is caused by the loss of some of the swollen granular bodies and the formation of entanglements between the polymers diffusing from the granules. It should be avoided in the production of starch pastes for smooth creamy products.

- The use of cross-linked starches, such as the chemically cross-linked distarch adipates, distarch phosphates, or distarch glycerols, provides swollen starches that resist this form of break down. They maintain a high viscosity and a short creamy texture for processing in boiling pans with simple stirrers.

- In more severe processing conditions involving heat exchangers, such as SSHE and canning processes, these starches may also fail. They need to be cross-linked more strongly and this means using the same types of starch, but having more cross-links present in the starting material.

- For other processes, such as in canning, the initial viscosity for suspending the vegetable or meat pieces is created with a native starch such as potato flour, and the heavily crosslinked starches added at the same time are not fully cooked out at temperatures up to about 80°C. Later in the process, as the temperatures rise in the cans to 110-115°C, the potato starch breaks down by diffusion and loses much of its viscosity, but the heavily cross-linked starch cooks out and takes over to give the final viscosity.

5.2.3 Setback and retrogradation

Cooked starch pastes are *meta*-stable and change with time, even when the granules have been cross-linked to protect them from overcooking. The starch polymers within the swollen starch granules tend to form double helices and recrystallise in sets of six as strong junction points. During this rearrangement they release some water from the hydrated chains. The effects on the swollen granules are that they gradually become firmer and denser as they lose control of their free water and it diffuses out of the swollen granules. This water may separate from the starch granules and seep out into the matrix of the food gel system in a phenomenon known as ***syneresis***.

The rate of change in starch gels increases at lower temperatures down to chill temperatures. If starch pastes are stored in a chill cabinet, their appearance and texture will change over a period of a few days, whereas at ambient temperatures they may take one or two weeks to show significant deterioration. Starch pastes become less translucent and their texture less smooth as they age, so that the product is noticeably different. If the products are frozen and thawed one or more times before consumption, the effects are accelerated and poor quality products ensue. In some cases where the products are reheated, the deterioration in quality can be reversed and the problem is not so serious. In other products that are eaten cold, such as fruit pies, custards and desserts, the loss of quality will become evident to the consumer.

The cooking process is very important in producing the most stable form of paste for any type of starch. It was shown that native starches must be cooked to above 90°C to ensure that all their residual crystalline structures are destroyed and that the optimum swelling is achieved (Guy and Sahi, 2006). This is difficult with unmodified native starches because they become very weak in their optimum swollen state and can easily be dispersed by mixers and pumping operations in large scale manufacturing processes.

The solution to this problem was found by the use of chemically cross-linked starch, such as distarch phosphate, distarch glycerol or distarch adipate with further mono-substitution by acetyl, phosphate, or hydroxypropyl groups. These special modified starches form more stable pastes when cooked and may prevent the loss of quality due to retrogradation during storage. The acetyl groups are less effective than the other groups and are best used for long ambient and chilled storage. Hydroxypropyl and phosphate substitution is superior to acetylation in all conditions and can give protection when the products are subjected to freezing and thawing processes.

5.3 Selection for a production process

The quality of the pastes is determined by choosing starch-based materials that can provide the final textural quality under the processing regime required for the product. In Figure 5.3 the range of processing conditions used to manufacture fluid products are shown with suitable choices for starch derivatives.

The selection of a starch derivative is determined by the severity of the processing conditions. For example, the following types of processing have increasing cooking time and shear input that tend to disperse the cooked starch granules unless they are strengthened by cross-linking.

5.4 Dry mix to be cooked for immediate consumption

For a dry mix product cooked up in the home, the main thickener would be a refined native flour or starch. It could be wheat flour, maize starch, potato flour or starch, or other native materials. The use of powders with coarse particle sizes >100μm, blended with the other ingredients, gives more protection against lump formation during water addition and a smoother product. Ideally, the particles should be <200μm for the best results.

Wheat flours prepared from steam-treated wheat, or which have been steam-treated themselves, have very low *alpha*-amylase levels and give thick creamy pastes at lower concentrations than untreated wheat flours (Guy and Sahi, 2006).

These flours are particularly good for products with longer cooking times and catering mixes that have to spend a longer time heating up than an instant

Starch performance - fluids

Figure 5.3: Selection diagram for starch derivatives in high moisture foods

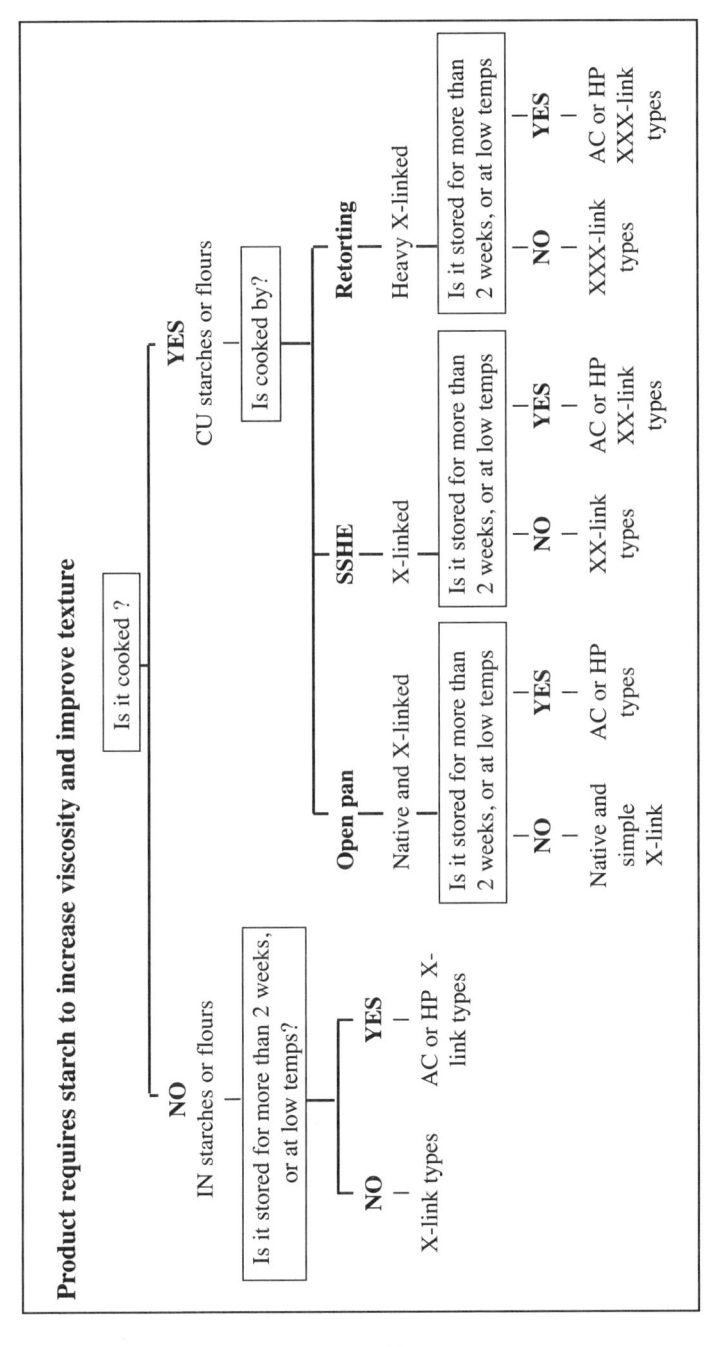

Product requires starch to increase viscosity and improve texture

Is it cooked?

- **NO** → IN starches or flours
  - Is it stored for more than 2 weeks, or at low temps?
    - **NO** → X-link types
    - **YES** → AC or HP X-link types

- **YES** → CU starches or flours
  - Is cooked by?
    - **Open pan** — Native and X-linked
      - Is it stored for more than 2 weeks, or at low temps
        - **NO** → Native and simple X-link
        - **YES** → AC or HP types
    - **SSHE** — X-linked
      - Is it stored for more than 2 weeks, or at low temps
        - **NO** → XX-link types
        - **YES** → AC or HP XX-link types
    - **Retorting** — Heavy X-linked
      - Is it stored for more than 2 weeks, or at low temps
        - **NO** → XXX-link types
        - **YES** → AC or HP XXX-link types

X-link light cross-linking
XX medium cross-linking
XXX heavy cross-linking

AC acetylated
HP hydroxypropyl

product cooked up at home. Such products might also be fortified with the addition of a small proportion of lightly cross-linked starch.

5.5 Instant starch added in cold processed sauce or dressing

Salad dressings, dips and cocktail sauces manufactured by low temperature processing must use an IN-starch type to give paste viscosity. The IN-starch has to be created by cooking quickly on a roller dryer or extrusion cooker and may need to be cross-linked to survive that cooking process with a good granular structure. There are some instant forms of native starches such as wheat and maize manufactured without chemical modification, but they do not form as smooth a texture in their paste in cold water as the IN-modified starches.

This is particularly true for the waxy maize starches, which are used for the excellent clarity of their pastes. They are easily dispersed on the roller dryers or extruders and must be protected by cross-linking to retain their short smooth creamy textures. Other starch types such as potato and tapioca are also used as cross-linked forms in this type of application.

The cross-linking used to protect the starches can also be introduced by thermal processing (Chapter 3) instead of the normal chemical methods. This gives a clean label product but little information is available at present on these products compared with conventional chemically modified starches.

IN-starch may be used at 3-5% to form a stiff paste in the sauce with the other materials such as plant fibres and the fat globules. The contribution of the non-starch components of the recipe may vary and also affect the viscosity of the starch. Fruit and vegetable fibres increase viscosity but fats and oils tend to decrease it. Any *alpha*-amylase present in other ingredients such as native flours, milk solids and herbs can have a devastating effect on texture during storage and must be eliminated.

As mentioned above (5.2.1), the IN-starch must dispersed carefully in water to avoid lumping of the powder.

5.6 Starch in recipes cooked in large pans to 85-90°C

The third type of process used in industry involves adding a cook-up starch derivative to a large boiling pan with steam injection and continuous stirring as the paste thickens. This process causes the starch granules to gelatinise and subjects them to shearing forces while they are swelling. The recipe may contain a mixture of native flours and starches and sufficient standard cross-linked CU-starch to create and maintain a particular viscous texture. The amount of time that the cooked fluid spends at high temperature after cooking determines the requirements for the cross-linked starch. If the fluid is held at 70-75°C for 1-2 hours, almost all the native flours and starches should be replaced by cross-linked starch to maintain the paste viscosity. Native starches break down to thin pastes under these conditions.

5.7 Starch in a recipe processed by Scraped Surface Heat Exchanger (SSHE) technology

In the SSHE units the starch becomes shear sensitive once it has gelatinised and swollen. Swollen granules are broken down by shear and by diffusion at high temperature so a fairly highly cross-linked starch is necessary to maintain the paste viscosity. In the SSHE units, the operation of the equipment is of primary importance to set the minimum shear input and to control the temperature so that the swelling of the starch occurs late in the transition through the unit.

5.8 Starch in a recipe processed by canning and retorting processes

For retorted products, a combination of starch types and pasting properties may be used, such as a combination of raw flour and some heavily cross-linked starch. Initially, as the product is prewarmed for filling, the native starches give sufficient viscosity to suspend pieces of meat or vegetable. At this stage, the heavily cross-linked starches used in the recipe do not swell enough to give more than a small fraction of the viscosity requirement to suspend the large chunks of material. Later, as the cans or pouches are transported through a retort at high temperature (110-120°C), the cooking shear and high temperature diffusion during the retorting will break down the native starches. In the final stages, the heavily cross-linked starch will swell and supply most of the viscosity. Usually a more heavily cross-linked starch than the standard material is used in these products. These materials provide viscosity and also help to keep the texture of the paste short and creamy.

Chapter 6

PERFORMANCE OF STARCH-BASED MATERIALS IN SEMI-MOIST SOFT FOODS

6.1 Overview

In this category of foods, which includes cooked cereals and tubers, pasta and noodles and baked cakes and breads, the starch is present in higher concentrations, in their slurries, batters or doughs, than in the fluid foods. The swelling starch granules raise the viscosity of the water phase to infinite values so that the foods become soft solids rather than fluids.

The range of foods in this category can be viewed in order of their cooked textures. These have been arranged in roughly increasing textural firmness values for the products shown in Table 6.1. Comparable values are not available for all the products so they have been listed in decreasing order of the amount of water available for the starch to absorb and swell.

The first band of products, which have the lowest starch levels and highest water to starch ratios, are flans and fruit pie fillings, where a soft gel is formed.

The second band are dry products that are cooked in excess water and absorb it at their surfaces, such as oat flakes, polenta, wheat bulgur, couscous, pearled barley and rice grains. They have increasing amounts of starch in the cooked product and decreasing water to starch ratios. The amount of water absorbed by the particles of grain is sufficient to gelatinise all the starch and give highly swollen granules. The final water to starch ratios are in the range 5-10:1.

The third band are products that are cooked in excess water but also contain a substantial amount of water in their structures, such as pasta, noodles, potatoes and other vegetables. The amount of water in potatoes (75%) is sufficient to gelatinise the starch without addition of extra water, as in fried chips and baked potato and Cornish pasties. Pasta and noodles have about 30% water, w/w, in their dough, which is insufficient to allow starch gelatinisation at <120°C. Therefore the diffusion of water into the dough is a key factor in their cooking process.

The fourth band of products are like the vegetables in that they contain sufficient water in their structures to gelatinise starch by heating to internal temperatures <100°C by baking, frying, or other techniques. Products such as cakes, samosas, spring rolls and the wide range of breads contain almost equal amounts of water and starch in their batters or doughs, and are cooked by baking, steaming or frying processes.

The swollen starch granules largely dominate the texture of this range of products. At the greatest water to starch ratios, the product texture is soft and weak, as in porridge, but as the ratio decreases, the starch granules are less swollen and become firmer. The firmest products are found in the last group of breads.

6.2 Fruit pie and flan fillings

The role of starch in tarts and flans is either to provide a gelling system on its own, or to support another gelling agent. These might be egg proteins, carrageenan/milk proteins, sodium alginate/milk proteins, or similar systems of other hydrocolloid gelling agents. Starch may be used in CU- or IN-forms depending on the process requirements. IN-starches can be useful to thicken a cold mix and keep particulate materials, such as fruit or vegetable pieces, suspended during holding, depositing and baking.

Fruit pie fillings

The role played by starch in a fruit pie filling is to thicken the acidic sugar syrup and to form a soft clear gel around the fruit pieces. An IN-starch can be mixed into some dry sugar and then blended into sugar syrup and mixed with the fruits. Its granules will swell to 70-80% of its final viscosity in the cold syrup and on baking will swell to its full paste viscosity. CU-starches will need to be suspended in the syrup, or added to hot syrup to thicken and give a stable gel.

Table 6.1: Semi-moist soft gel products

	Starch range %	Starch %	Water %	Other ingreds %	Ratio water/ starch	Ratio starch/ water
Flan /custard	5-7	6	80	13	13.3	0.08
Fruit pie filling	4-7	5	50	50	10.0	0.10
Oatmeal porridge*	9-11	10	85	5	8.5	0.12
Maize polenta*	12-15	14	85	4	6.1	0.18
Bulgur*	18-22	20	78	2	3.9	0.26
Couscous*	22-26	24	74	2	3.2	0.31
Pearled barley*	25-28	25	73	2	2.9	0.35
Rice*	30-32	30	66	4	2.2	0.46
Noodles*	13-15	14	84	2	6.0	0.18
Potato*	14-18	20	77.5	2.5	3.9	0.26
Pasta*	18-22	22	76	2	3.5	0.29
Potato fried	29-30	30	56	12	1.9	0.54
Meat pastes	8-12	10	67	23	6.7	0.15
Sausages	20-25	25	50	25	2.0	0.50
Cakes high ratio	24-26	25	30	45	1.2	0.83
Cakes low ratio	30-32	31	27.5	41.5	0.9	1.11
UK bread	47-48	46	45	9	1.0	1.00
Pita bread	46-48	46	40	14	0.87	1.15
Samosas	48-50	48	42	10	0.88	1.14
Spring rolls	43-44	42	52	6	1.23	0.81

* cooked in water

Native starches or flours will be unsatisfactory, either in IN- or CU-forms, because they will break down and become cloudy on storage due to retrogradation. The best forms of starch for use in fruit fillings are the modified waxy maize, potato or tapioca starches. They should be cross-linked to prevent significant break down and substituted with acetyl groups, monophosphate or hydroxypropyl groups to inhibit retrogradation. The processing of fruit pie fillings can vary from boiling pans to canning processes, and therefore the amount of cross-linking in the starches can vary from light to heavy. Similarly, the storage of the filling may vary from ambient to frozen conditions, and the substitution may need to change from acetyl to hydroxypropyl or phosphate groups. Their gels made with 3-5% starch in the syrup will be translucent and remain stable for several weeks of ambient storage, or several freeze-thaw cycles.

Other gelling agents such as alginates, pectins or carrageeans may be used with the starch to give a stronger, more elastic gel texture. Normal modified starches do not interfere with these systems but give their normal thickening effect within the gel systems.

Flan fillings

Savoury flan fillings are made with combinations of starches and egg proteins to give a soft gelled texture after baking. There may be other gelling systems in the recipe such as milk proteins and hydrocolloids, but the starch will give a good thickening effect in all other gel systems. Normally these fillings are kept in chill store for a short shelf life, or they may be frozen, or they may be mixed as batter, frozen and baked from the frozen state. The modified starches are needed in all these types of products except those consumed within a few hours of baking. Any retrogradation, or too much dispersal, of the starch can cause serious faults in the appearance and texture of the flan filling. The modified cross-linked starches from waxy maize or tapioca, with some secondary substitution by acetyl or hydroxypropyl groups, are the preferred types. An important factor for any starch that is used is that it should be cooked to its optimum paste and not be undercooked.

6.3 Cooked grains and tubers

The simplest products in this category are the cooked tubers, such as potato, cassava and sweet potato. These raw materials contain their starch in plant cells with a large supply of water, roughly three times their starch content. The tubers, or cut pieces of vegetable, can be easily heated to above the T_m values of their starches (55-70°C) to allow the granules to hydrate and swell. This change will transform the tuber from hard firm bodies to soft deformable structures.

In several applications, the tubers will be cooked in boiling water to supply the heat and will also be able to absorb water in the outer layers. There will be a texture transition from the edge of the tuber to the centre and overcooking will lead to excessive softness and cook out of starch into the water.

In other products, such as pasties, pies and potato chips, the only water available will be that present in the raw material. The cooking process must be designed to heat the product to above T_m in the time available, using the frying or baking conditions to obtain a soft cooked product.

Cooked grains do not contain enough moisture to gelatinise their starch granules unless they are heated to >150°C as in extrusion cooking or UHP cooking. The cooking process requires that extra water is made available to diffuse into the grain or particle. The cooked cereal products might be polished rice, pearled barley, oat flakes, or prepared wheat derivatives, such as semolina, wheat bulgar and couscous. In all these products, the hot water provides both the moisture to obtain a T_m <100°C and the heat to raise the grain temperature. Gelatinisation of the starch occurs first at the edges of the products and moves inwards to the centre with cooking time. Once again, there is a gradient in firmness of the product from the edge to the centre and some of the starch at the surface diffuses outwards into the cooking water.

A view of the fine structural changes taking place during the cooking of three cultivars of Indica rice was given by Takahashi *et al.* (2001). They observed the cooking of milled rice grain in a rice cooker with a scanning electron microscope. After cooking rice for 15min, the starch granules in the first cell layer of the rice grains became swollen and a fibrous structure developed on their surfaces. The swelling starch granules in each amyloplast cell became fused together in a solid mass after about 20min cooking, and then developed into an irregular gel-like structure. The rate of swelling of starch granules and the formation of the irregular gel-like structures during cooking processes were slower in the inner layers and at the central part of a grain.

The final moisture levels in the cooked grain and tuber products are in the range 72-76% w/w for rice and pearled barley, but are higher for cooked maize polenta, oatmeal and boiled potato.

6.4 Pasta

6.4.1 Traditional pasta

Pasta is manufactured by mixing dough and forming fresh pasta at about 32-34% moisture w/w, either by extruding, or sheeting and laminating it out with pairs of rolls. The fresh pasta can be dried slowly to manufacture traditional long-life dried pasta products at about 5% moisture w/w. At this stage, the starch is unchanged from its native state but the proteins have been developed into a continuous phase.

Wheat semolina forms the basis of all pasta and may be enriched with eggs to form egg pasta. The normal semolina for authentic Italian pasta is milled from *T. durum* wheats, but in several countries *T. aestivum* hard wheat semolina is used instead for economic reasons. High quality durum wheat has a vitreous kernel with a high protein content, good gluten forming ability and a clear bright yellow colour.

The choice of raw materials is restricted to durum wheat in Italy, France and Greece, but other countries allow both durum and other types of wheat grist to be used in part or as a complete replacement for durum wheat in pasta. The best alternative wheats are hard endosperm breadmaking wheats from *T. aestivum*. The USA and Italy are by far the largest consumers of pasta and in Eastern Europe and Asia most of the pasta is made from the other wheats.

The new waxy durum wheats have been tested in pasta and found to be less satisfactory than normal durum. They tend to swell too much and give a softer texture and more cook out (Gianbelli *et al.* 2005; Vignaux *et al.* 2005).

Manufacturing

The semolina is mixed with water at about 30-32% moisture w/w and allowed to hydrate. After 20-30min, the moist semolina is extruded at 40-45°C to form dough in which the soft hydrated protein is developed into a continuous phase around the starch. Finally, the dough is extruded from dies into fine strands, sheets or ropes that can be cut to manufacture products such as spaghetti, lasagne and various geometric forms of short cut pasta. Traditional semolina has a particle size range of 200-450μm, but in some new processes a smaller size of semolina, with 130-350μm particle size, is used for faster hydration during processing. The pasta may be dried, partially cooked, or fully cooked from the raw dough.

Dried pasta: Drying processes have been based on low temperatures of 40-60°C for pasta made from durum semolina and higher temperatures (HT) of 60-84°C (Zweifel *et al*. 2003) or even higher (UHT) at >84°C for pasta made from bread wheat semolina. The high temperatures serve to modify the protein phase surrounding the starch and make it more resistant to cooking.

Fresh pasta: The extruded pasta may be chilled and sold, or used in that form, preferably after a heat treatment to destroy enzymes and micro-organisms. This is done by passing the pasta through a steam chamber, or hot water bath, to raise its temperature to about 88-90°C for 1-2 min. This leaves most of the starch ungelatinised because the moisture level is too low, but cereal *alpha*-amylases are destroyed.

Partially cooked pasta: Pasta can be passed slowly through a tank of boiling water to partially gelatinise the starch within its structure and increase its moisture content to 40-45% w/w. This may be done with sheets of lasagne, or ropes of spaghetti, to give a good cooking performance for the final bake-off of frozen ready meals.

The final cooking process for pasta is based on heating the dough or dried pasta in an excess of boiling water for a set time to produce a cooked texture, called *"al dente"*. This texture may be described for a piece of pasta as varying from a soft outer zone to a firm centre.

Pasta is cooked in an excess of water, but initially only the outer layers of the products have contact with the extra water. Their starch granules are able to gelatinise, but a short distance from the surface all the dough is at 30% water content, where the gelatinisation temperature, T_m, is >100°C and consequently at first its starch cannot gelatinise. The cooking process proceeds from the outer edge as the starch granules swell and draw water into the pasta. The dough temperature rises and the T_m of its starch granules falls to 55-60°C due to the increase in water content. The changes to starch were shown by light microscopy and a SEM technique to vary from a hydration-driven gelatinisation process in the outer layers, to a heat-induced crystallite melting in the central layers. Once melted, the granules are able to draw water in from the outer layers and form a moisture gradient from the surface to the centre of 90 to 39g/100g. The cooking interface moves into the pasta pieces with time until all the starch is gelatinised (Conde-Petit *et al.* 1998). However, the amount of swelling of the starch granules will be low at the centre and will increase to the maximum amount at the edge of the pieces. This can be clearly demonstrated for spaghetti, with central starch having a firm texture and the surface being soft and losing some starch (5-6%) into the cooking water.

The cooking quality of the pasta is judged both by the texture of the cooked pasta and the amount of solubles in the water. Good quality *T. durum* semolina will require a longer cooking time and release less soluble material than poor quality semolina. Microscopy reveals that a strong gluten network is formed in good durum wheat dough, which surrounds the starch granules. This protein network slows the cooking process by slowing the diffusion of water to the starch, but it also protects the starch granules from being overcooked. The protective gluten network also reduces the diffusion of starch into the cooking water. The differences in cooking quality between the dried pasta and the fresh pasta are fairly small, if they are both made from the same quality of semolina.

T. aestivum semolina gives poor quality fresh pasta, but it can be improved when used to manufacture dried pasta by drying at relatively high temperatures of 70-80°C (Milatovic and Mondelli, 1991). This increases the strength of the protein network to give a better match to that of *T. durum* semolina pasta. Reports also suggest that adding transglutaminase enzyme (Kovacs, 2005) to cross-link the proteins also gives a significant improvement in texture and less cooking loss.

The addition of egg to all types of pasta helps to form a better protein barrier to prevent starch cook out losses and give a better texture. The minimum level for egg pasta in Italy is 20% liquid egg in the pasta, or about 5% whole egg solids. The albumin mixes well with the wheat proteins and forms a strong supporting gel network to help control the gelatinising starch.

Storage of cooked pasta - changes in texture

If pasta is cooked and then stored, as in chill or canned products, it will change with time. The "*al dente*" central zone of firm texture will disappear. The gradient of partially swollen starch leading from the surface to the centre of the pasta will inevitably draw moisture from the edges of the pasta inwards to the centre. Once the starch granule loses its crystallinity it has a potential to absorb water and expand. There will be an equilibration of moisture across the pasta that will result in the texture becoming softer in the centre and loss of the "*al dente*" quality. Undercooking the pasta to leave the starch in the central zone ungelatinised can leave a better texture on reheating the pasta.

If there is a source of moisture available, as in a sauce that is in contact with the pasta, that moisture will also be drawn into the pasta. The gluten network and the presence of gelled egg proteins reduce both the rate of moisture flow into pasta and its softening.

6.4.2 Pasta from unconventional raw materials

Pasta is an attractive and convenient food and is most easily made from the semolina of *T. durum* or hard bread wheats. However, similar products are made from other cereals such as maize, rice and barley. The derivatives of these cereals are not suitable for use in a conventional pasta process because their proteins cannot form a protective network around starch during cooking.

In developing pasta from such products, the technology has to be changed to improve the dough structure. Traditional rice pasta products are made in the

Far East by gelatinising a portion of the starch prior to extrusion to bind the flour particles together with gelled material.

In a typical process, rice flour is enriched with legume flours, such as mung bean flour, to improve nutrition. After adding water to the mixture, the dough is extruded and heated to 90-95°C to gelatinise some starch. The cooked rice pasta is immersed in cold water to form a stronger structure, and reheated to 90-95°C one or more times to increase the texturisation, before finally the pasta is dried in the same way as conventional pasta.

The mechanism for the texturisation of rice pasta is based on the crystallisation of amylose. Some of the starch granules are gelatinised and swollen during cooking to disrupt their structure and form an exudate rich in amylose. The amylose present in the exudate crystallises quickly on cooling and builds networks and junction points throughout the strands or sheets of pasta. Crystalline structures formed by amylose do not remelt during reheating or subsequent cooking in water to prepare the pasta products. Successive heating and cooling cycles for the pasta increase the amount of amylose crystallisation and improve the final cooking performance of the pasta. The selection of rice with high amylose levels (25-30% w/w) is beneficial in these products.

The traditional process has been developed for industrial use by cooking 5% of the starch-rich material in excess water (1:5), blending it with the uncooked fraction and extruding to form the pasta. Adding proteins such as egg and whey proteins can also help to stabilise the structure and improve nutrition and flavour. Buhler (2005) has developed an industrial process to upgrade its existing pasta lines to produce pasta from a number of gluten-free starch-rich materials such as maize, rice, potato and palm starches.

6.5 Noodles

Noodles are a major food product in the Far East and sales are growing in Europe and North America. They are made in a number of forms from good quality *T.aestivum* flours (Kruger, 1996) in manufacturing processes closely resembling pasta production. The wheat flour is mixed with water to give a crumbly dough of 32-36% moisture, w/w. Other ingredients may be added in small quantities, such as sodium chloride and in some cases other alkaline salts (Shiau and Yeh, 2001). The low moisture dough is crumbly after mixing but can be formed into a true viscoelastic dough, either by low temperature extrusion or sheeting on pairs of stainless steel rolls, and cut to form the final products. Most current processes use five pairs of sheeting rolls, forming the dough into thin sheets about 2mm thick. The process uses four pairs of reduction rolls, before cutting the sheet into thin strips with a pair of cutting rolls. The strips are cut to length and are usually steamed for 1-2min, or passed through boiling water for 30-60s to pasteurise them and remove enzyme activity. There is little gelatinisation of the starch in the processing and all the product needs to be cooked in boiling water before consumption.

The strips of dough can be cooked in boiling water in a few minutes and consumed, chilled and stored, or fried and stored. The final cooking process in boiling water causes all the starch to be gelatinised and the noodles to become soft, but like pasta there is a textural quality that is recognised by the consumers. For optimum cooking, there should be a firmer texture in the central region of the noodles.

Differences were noted between noodles made with wheat containing normal starch and waxy starch. The waxy starch cooked more easily and swelled more extensively in the water to give a softer noodle. This was beneficial in making Udon noodles, which have a special soft unctuous texture and are an important type of noodle in Japan.

Cooked noodles are sold both as fresh cooked and long life products. The same problems concerning loss of texture occurs in the cooked noodles as in cooked pasta due to the equilibration of the moisture from the outer layers to the centre on standing for a few hours.

6.6 Cakes

Flour confectionery products such as cakes have recipes that contain large amounts of sugar and are formed from liquid batters. The sugar dissolves in the available water during mixing to form concentrated sugar syrup (45-50%w/w). This syrup surrounds and permeates into the starch granules, and changes their gelatinisation temperature. As the starch granules gelatinise and swell in the syrup, the solid cell wall structure of the cake is created and cools in a meta-stable form. The role of the starch will be considered in two stages: the creation of the cake structure, and the changes occurring during storage (the staling of the cake).

6.6.1 General cake recipes

In normal cake recipes for either high- (HR) or low- (LR) ratio cakes, the basic raw materials are flours prepared from soft wheats. Cake flours have more free starch granules and fewer mill damaged granules than bread flour. Special milling techniques, using turbo milling with air-classification, can be used to produce superior cake flours with large proportions of free granules and an average particle size <50μm.

Starch granules absorb about half their weight of syrup and then remain inert in the batter during the early stages of baking. The normal gelatinisation temperature in pure water for wheat starch, assisted by the action of water molecules on the polymer chains leading into the crystallites (Section 1.8), is about 55°C. However, the presence of 45-50% sucrose in the water reduces the kinetic impact of water molecules on the amorphous polymer chains leading into the starch crystallites. Sucrose and other small carbohydrates can control a shell of water molecules around their own molecules. This reduces the average kinetic energy in free water molecules and it requires more energy in the aqueous phase to melt the crystallites. Therefore, the temperature of the batter has to be increased to melt the starch crystallites. The effect of sucrose increases with concentration and at the levels used in cake syrups the gelatinisation temperatures are raised to 85 to 90°C.

Other common carbohydrates used in flour confectionery have similar effects to sucrose (Bean *et al.* 1978), but as the molecular size decreases from sucrose to glucose, xylose and glycerol, the elevation of the gelatinisation temperature also decreases. This is because their control on water molecules is less powerful than sucrose, roughly in the order of their molecular size and number of hydroxyl groups.

The effect of small carbohydrates is similar for all types of starches that might be used in a cake batter. It tends to increase the gelatinisation temperature from that found in pure water in proportion to the concentration of carbohydrate used in the syrup. Starches such as maize and rice that have higher gelatinisation temperatures than wheat starch, will maintain their differential as the sugar concentration is increased.

6.6.2 High- (HR) and low- (LR) ratio cakes

The melting of starch crystallites and swelling of starch granules rapidly increases viscosity within the foam structure of cake batter, concentrating the egg and flour proteins between the swelling granules. The swelling of the starch granules solidifies the structure, stops gas cell expansion and causes the gas bubbles to rupture and release their excess pressure. The change from foam to open sponge structure moves inwards from the outer layers of batter to the centre of the cake. The final stage of solidification occurs as the egg proteins gel in the intergranular spaces.

In low ratio cakes, flour is used with 45-50% sugar syrup in proportions of 1:1.5 to form a starch gel within the cake crumb at 85-87°C. This gives a moist crumb with a soft elastic texture in the cake that stales significantly over two to three weeks.

High-ratio cakes are made to give a lighter more melting texture by reducing the flour level by about 20%. The ratio of starch to syrup in the cake gel falls to 1:2.2 and at these levels untreated cake flour fails to stabilise the sponge structure, which leads to extensive collapse of the whole cake structure. The gel strength achieved by untreated flour during structure formation in LR cakes must be achieved in the HR cakes with 20% less flour or else the structure would be weak and the gas cells in the centre of the cake would continue to expand and not rupture during baking. In this scenario, as the water vapour in the cells condenses, they collapse completely, forming dense areas in the sponge and causing massive collapse to the whole structure of the cakes.

The reduction in starch level was only made possible by applying special treatments to the wheat flour to improve its starch performance, particularly its ability to form a strong gel in the cake syrup at the lower level (Guy and Pithawala, 1981). At the lower concentrations of starch used in HR cakes, the viscosity of the batter made with a treated flour increases to the level required to stop the expansion and cause the cells to rupture.

This increase in viscosity was attributed to either,

- improved swelling of starch granules to give stronger intergranular contacts and greater concentration of the gelling egg proteins in the intergranular spaces
- or an increase in the release of soluble starch into the intergranular space to strengthen inter-granular links,
- or a combination of both.

The performance of the starch was improved by the following flour treatments: chlorination and dry heat-treatment at >120°C. After these treatments are applied, cake flour has greater viscosity and gel strength when heated both in water and syrups with high sugar levels (Frazier *et al.*, 1974) and magnifies the effect on gel strength due to egg protein (Guy and Pithawala, 1981). Currently only heat treatment is permitted in the EU although chlorination is still used in the USA. The gelatinisation temperature of the starch is not changed significantly by the optimum treatment and the swelling of granules begins at the same point in the baking cycle. The improvement in starch performance is in the dynamic increase in viscosity to a high level to form a solid cake structure by the end of baking.

6.6.3 Interactions with other materials

The creation of a starch gel in sugar syrup is the basis of cake structure but the interaction of starch granules with egg proteins is also very important. In the critical phase of cake structure creation, the starch granules gelatinise and swell, but at the same time the egg albumin proteins start to polymerise and form gels. The swelling phenomenon of starch granules is a much faster process than the gelling of egg proteins and therefore the swelling granules concentrate the egg proteins in the intergranular spaces in a dynamic expansion. Egg proteins, unlike small sugars, are too large to enter the amorphous regions of the starch granules. As the bake continues, the egg proteins polymerise under a temperature-time relationship to form an ordered cross-linked gel structure around the starch granules. The overall structure of the cake cell wall has been described as a brick wall in which the starch granules are the bricks and the egg gel is the mortar. Egg albumen's components, ovalbumen and conalbumen, make the main contribution of all the egg proteins to the textural properties of the mortar.

6.6.4 Flour confectionery: retrogradation of starch and crumb firming

The study of cake staling phenomena revealed that the consumer's judgement of staling was determined by two major sensory attributes: crumb firmness and moistness (Guy *et al.*, 1983). Control of moisture loss and movement within cake products can reduce staling considerably, but there still appears to be an intrinsic form of staling in the crumb (Chamberlain, 1962; Guy, 1983).

Studies in cakes and model gel systems showed that wheat starch forms a soft gel in aqueous sugar syrup, which like bread crumb firms to a plateau value after several days. The firming process for the cake starch gel can be reversed by heating to a high temperature, but on cooling to ambient temperatures, the gel begins to firm again. This firming phenomenon shows a maximum rate at 10-15°C, indicative of a polymer gelling mechanism. However, there was no evidence for the formation of crystalline structures in the ageing gels, either by DSC or X-rays. It is possible that the reversible firming effect is related to the formation and remelting of starch double helices.

Studies in starch gels showed that the reversible gelling phenomenon is similar to that found in bread (Guy, 1983). However, the refreshening temperature of cake crumb was much higher than bread in the range 90-105°C; consequently it is not easy to refreshen cake. The anti-firming emulsifier additives used to inhibit bread staling were not very effective in cakes (Guy *et al.*, 1983). Some reduction in firming rate has been observed with maltogenic amylase in cake donuts and sponge cakes but the results in HR pound cakes were not of commercial benefit (Guy and Anstis, 2001).

6.7 Standard UK bread products

6.7.1 Structure formation

As in cakes, starch plays two roles in bread products, firstly in the creation of the sponge structure and secondly in the changes to the stored products, known as staling. Therefore, this section is divided into two parts to consider these aspects separately.

The gluten forming proteins of wheat determine the gas holding quality of the foam created in bread dough. However, the starch granules, present at about 75% of the solid material in the gas cell-walls, also play an important role in spreading out the gluten into thin films to contain bubbles of air, during dough mixing. Later in the baking process, they swell to stop the expansion of gas bubbles, causing a pressure increase that eventually ruptures the gas cells to form the sponge structure of breadcrumb (Guy, 1995).

Wheat starch is well suited to these functions as its gelatinisation and swelling occur at a critical point in the expansion (60-75°C). The basic raw materials prepared for breadmaking contain starch granules partly embedded in a protein matrix (wedge protein). Some recent studies have focussed on the effects of starch granules size (Sahlstrom *et al.*, 2003a;b) and it is necessary to consider this variable.

There are two distinct sizes of starch granules in wheat flour, with proportions based on mass of

- 80-85% of the larger A-type granules (15-40μm) and
- 15-20% of the smaller B-type granules (1-10μm)

Two studies have been carried out to evaluate the effects of A- and B-type starch granules. In the first study granules were separated from 6 different bread wheat flours of similar protein contents and analysed by DSC and RVA. The resulting data was modelled for variations in the bread characteristics of mass, volume and shape. There were significant differences in the results for each starch type. A-granules had the lowest gelatinisation onset (T_m) and peak temperature (Tp) and the largest melting enthalpy/g for the main crystallites in the granules. B-type granules had the lowest onset temperature for amylose/lipid complexes and highest melting enthalpy/g for the complexes.

RVA analysis under standard pasting conditions showed that the A-granule fraction gave the highest peak viscosity, break down, setback and final viscosity for a fixed mass.

The analysis of the resulting data gave a correlation between some of the A-granule's DSC and RVA variations and the mass of the loaves, but showed no significant relationship with volume or shape. The optimum values for the A-granules gave the greatest retention of moisture and the highest mass.

The second set of studies, based on the performance of a range of flours in 'pup' loaves (Park *et al.*, 2004), showed that optimum levels of both types of starch granules are needed to give the best loaf mass and volume. The larger A-type granules have the most important role, but the smaller B-type granules are also important. A positive linear correlation was found between the crumb grain scores and the size distribution of the A-type granules, and a polynomial relationship occurred between the crumb scores and percentage of B-type granules. The best crumb grain score was obtained with flour containing 19.8–22.5% of total starch as B-starch.

In well-mixed dough, the intimate contact between starch granules and the surrounding protein gel phase raises the gelatinisation point of native wheat starch from 55 to about 62-64°C. The swelling starch granules increase dough viscosity and terminate the extension of the gas cell walls that is found during oven spring. Rising gas pressure ruptures the gas cells and forms the sponge structure and equilibrates the crumb with atmospheric pressure. The final loaf volume is stabilised by the increasing viscosity of the starch paste in the crumb as it cools.

The oven spring process can be extended by the action of *alpha*-amylases, which are active at high temperatures. They hydrolyse some of the outer starch layers in the swelling starch granules to reduce viscosity in gas cell walls for a short time and allow greater expansion of gas cells to increase loaf volume. Cereal *alpha*-amylases are active up to 85°C and perform this improvement but if their levels are too high, they seriously weaken the crumb by reducing starch viscosity and elasticity too much. The breadcrumb becomes weak and sticky and cannot be sliced. Fungal *alpha*-amylases are more easily denatured by about 75°C (Pritchard, 1992) and have less time to act on the starch. They can be added to supplement the cereal *alpha*-amylases and give better control of the increase in loaf volume while maintaining good texture in the crumb.

After baking, the swollen native starch granules retain a high proportion of the water added to the dough (particularly the A-type granules). This helps to give a soft texture to the baked crumb and attempts have been made to increase the moisture content of breadcrumb by adding other starches. The addition of mill-damaged starches was found to increase dough moisture but little of the extra water was retained. The damaged starches were either degraded by the *alpha*-amylases present, or had insufficient structure to hold the water at high temperature.

New forms of wheat starch with high amylopectin levels, such as waxy maize, have been tested in bread. They are weaker than normal wheat starch after gelatinisation and can break down in the cell walls of the dough and give poor crumb texture at high levels of usage. However, at low levels up to 20-30% they combine with normal wheat starch to give a good crumb structure with a softer texture, due to extra moisture retention.

The use of chemically modified starches such as starch monophosphates at 1-2% has been more successful in retaining water, but are not normally permitted in many countries.

6.7.2 Starch in bread crust

Bread can be baked, with or without steam injection, to give different crust characteristics. In a dry oven system the dough loses water rapidly at the surface, and before the starch granules can gelatinise the moisture level is too low. A few millimetres below the surface the moisture level rises to 40-45% w/w and the starch granules are able to gelatinise. Thus, the surface crust has a layer of ungelatinised starch which forms a rough texture. The crust containing ungelatinised starch has a slightly rough appearance.

If dough is placed in the same hot oven with steam injection, a remarkable change takes place and the baked bread develops a smooth glossy crust. Steam condenses on the cool dough entering the oven and raises its temperature very quickly to >80°C. Starch granules in the surface have enough moisture to gelatinise at 65-70°C and form a soft gel. As the dough expands, the gel stretches to follow the expansion and as the temperature rises to 100°C and water evaporates, starch polymer forms dried films to give a glossy appearance to the crust. Manipulation of the steam treatment can cause variations in the glossy crust from a thin elastic shiny layer to a thicker layer that cracks on cooling (Wiggins, 1998).

6.7.3 Bread products: retrogradation of starch and crumb firming

The swollen starch granules within wheat, gluten-free and rye breads, and similar moist products are intrinsically unstable and begin to form double helices and recrystallise after the products cool to below about 50°C. The formation of crystalline junction points between molecules across the gelled starch inside the granules causes them to become firmer. Consequently the breadcrumb itself becomes firmer and the products are said to stale.

The use of chemically modified starch is not permitted in bread and would probably be too expensive, so some other approaches have been developed to deal with the problem.

Lipid complexes between amylose and monoglycerides

In the first methods developed to inhibit staling, surfactant lipid molecules with long saturated fatty acid chains (C16 and C18) were used to form complexes with the starch polymers. These lipids had to be formed into a micellar structure with water (Krog, 1971) before addition to the bread dough to be effective. Individual molecules could split from the micelles and complex with amylose and possibly with amylopectin inside the starch granules to slow the rate of recrystallisation of the amylopectin. They did not change the initial or final firmness value of the bread very much, only the time taken for the firmness to reach its plateau value. High levels of monoglyceride >0.5% of the flour weight are required for significant effects and they may reduce crumb elasticity. This fault can be remedied by adding more gluten, or 0.3% DATEM ester on flour weight. This gives 2-3 days extra life to bread before it is considered to be unacceptably stale by consumers.

Starch modification by maltogenic amylase

The second approach developed in the 1990s uses an *alpha*-amylase enzyme to modify the starch molecules within the granules while the bread is baking (Outtrup, 1986; Outtrup and Norman, 1984). After gelatinisation and swelling in water at 65-75°C, starch granules are susceptible to hydrolytic degradation by heat-stable amylases until the enzymes are denatured and become inactive. The cereal and fungal *alpha*-amylases attack starch randomly and reduce its viscosity significantly. The new form of *alpha*-amylase has an *exo-* action on the amylopectin side chains, removing maltose units in a slower reaction. It has a limited time to act because it is destroyed as the temperature in the dough reaches about 85°C. The changes to the swollen starch granules are subtle in that the firmness of the freshly gelled starch is not reduced. However, the shorter polymer chains do not recrystallise to form as many crystalline junction points during storage. The rate of firming of the starch in bread is not reduced very much but the final plateau value for the firmness is greatly reduced, so that the bread remains fairly soft almost indefinitely, if it is in hermetically sealed packaging.

6.7.4 Bread without gluten

A small part of the population in the EU and other areas of the world suffer from gluten intolerance and cannot eat products manufactured from wheat flour. This creates a problem in supplying bread, or a bread-like product. The use of separated wheat starch, barley or rye flours may introduce small amounts of wheat and similar type proteins and must be avoided. Commercially available starches from rice, tapioca, maize and potato have been tested and used in gluten-free breads but are all are less satisfactory than wheat starch.

- potato starch gelatinises early and restricts oven rise
- rice gelatinises late, has small granules and gives a weak cell structure
- maize starch gelatinises late but its granules are the right size
- tapioca and waxy maize have weak granular structures and give collapsed areas of crumb
- all the starches lack gluten proteins to hold gas.

The best solutions to the problem appear to be to use combinations of rice or maize and tapioca or potato with a cellulose-based or natural hydrocolloid gum, to form a batter, from which bread-like products can be baked (McCarthy *et al.* 2005). Skim milk powder and other proteins can be added to give a good nutritional profile because the hydrocolloids can carry them in their continuous phase. The viscous thermo-setting cellulose ethers and xanthan gum have been reported as the best substitutes for gluten. Recent research with transglutaminase enzymes has shown significant improvements in forming doughs from rice flour and baking bread from them with a good volume.

6.7.5 Breads with rye flour

The role of starch in rye bread is similar to its role in wheaten breads, but because the rye protein forms only a weak gluten, the continuous phase of the dough is developed from the hydrated hemicellulose fraction. It is like the gluten-free bread in that hydrocolloid gums form a thick paste around the starch. Gelatinisation and swelling of the starch granules of rye causes the change in structure from foam to sponge as in wheaten bread. However, the gas holding power of the rye dough is low and the structure is dense. Therefore, rye flour is blended with wheat bread flour at levels up to about 50%w/w to give a range of products with increasing loaf volume and lower crumb density, while retaining the unique rye flavour.

6.8 Naan breads

Naan breads are manufactured from a recipe of flour, water, yoghurt and minor ingredients. An example of a manufacturing process would involve mixing a dough in a spiral mixer for 2min slow and 8min fast speeds at 26-28°C. The dough is sheeted down to 12mm and turned through 90° before gradually sheeting down to 3mm. The naans are cut from the sheets, docked, proved at 30°C for 45min and baked at 300°C for about 2min. The starch granules gelatinise to give a soft texture, which stales in the same way as bread and is sensitive to moisture level and loss.

6.9 Pita bread

Pita bread is made from strong bread flour with a recipe containing normal bread ingredients, fat, salt and yeast. Dough is mixed to 28°C at 2min on slow and 7min on fast speeds on a spiral mixer and fermented for 30 min at 27°C. It is sheeted to 12mm, turned and sheeted down to 2.5mm and cut into discs of 15cm diameter. The doughs are proved in a dry atmosphere for 20min and baked in an impingement oven set at about 230°C for 3min. The starch granules are gelatinised and give a soft texture except at the surfaces, where the moisture is reduced to form a dry crust.

6.10 Samosas

Samosa dough is mixed from flour and water (1:0.4) with small amounts of oil and salt. It is blocked out into a square chunk and allowed to rest for 1-2 hours on racks. Some lamination is done at this stage but only once or twice. After resting, the dough is sheeted quickly down to a half thickness and then two sheets are added together and rolled out, folded, laminated several times and reduced to 1-2mm by 6-8 passages through a pair of rollers. The sheet is then passed through a small gap on heated parallel bands at about 300°C to form the final sheet with a surface layer of gelatinised starch. After rolling it is cut, coated with edible gum adhesive, filled with meat or vegetables and folded to form samosas. These are fried at 180°C for 2-3min. Most of the starch in the dough is gelatinised and the dough is soft except for a thin crisp layer on the surface and has characteristic surface blisters. The air layers created by lamination remain in the final sheets to form bubbles that are still present as the water is boiled off in the frying unit.

6.11 Spring rolls

Spring rolls are made from a fluid batter of wheat flour (40-45 % w/w) in water with some salt and minor ingredients (120% water, fw). The batter is deposited onto a heated roll and the cooked sheet of "paper" removed on scraping knives. The flour contains native starch, damaged starch and protein (roughly in proportions of 65-70%, 5-7% and 10-12%, respectively). After mixing in water, the wheat flour proteins and the damaged starch hydrate and swell, giving the cold paste viscosity. Any amylase in the flour acts to break down the damaged starch, producing reducing sugars and reducing the paste viscosity.

The wheat flour slurry is deposited into the feeder system of the roller dryer. This consists of small doctor rolls that dip into small tanks of batter and then deposit a thin layer on the hot main cooking roll (temperatures 110 - 190°C). A thickness guide that can be screwed in to change the thickness of the layer on the smaller roll controls the pick-up of wheat flour slurry. At the surface of the larger roll, a strip of wet wheat flour slurry is laid down and meets the hot roll surface at temperatures from 100 to 110°C. The roll rotates at a fixed speed under an extraction cover to reach the scrapers, where a sheet of moist cooked wheat flour pastry is removed and flows onto a conveyor belt in a continuous stream.

The wheat flour pastry sheet, which should be stretchy and smooth without imperfections such as holes, bubbles or rippled surfaces, is fed into the wrapping machine where small cylinders of filling are deposited onto the sheet before it is folded. The wrapping continues with folding, cutting and then rolling up into a Swiss roll style package. The spring rolls are then fed into a continuous frying unit where they are fried for 3-4min at 180-190°C and become slightly brown and crisp on the outside, but most of the pastry remains at 35-40% moisture.

Traditional spring rolls in the Far East are made from rice flour by a labour intensive industrial method involving the steaming of thin layers of flour on fine sieves. The manufacture of pastry by modern hot roll processing is also being developed and it should be possible to use other cereals to form similar products.

6.12 Cooked meat products

Cooked meat pastes need water binders to give a soft gelled texture that is easily spreadable for products such as potted pastes and pates. For more solid products like frankfurters, meat puddings and sausages the starch helps to partially solidify the structure before cooking completes the solidification by gelling meat proteins.

The coarsest textured materials are special forms of gelatinised crumb and are prepared from bread and biscuit type doughs. These absorb water and form the solid texture in sausages and meat loaf type products.

Finer textured water binders are the potato flakes and granules (Chapter 3A.3) that consist of gelatinised potato starch and cell wall materials. The flakes have a majority of free granules and give a fine texture similar to potato starch.

The finest textures can be formed from starches and gums, and these can be blended into meat emulsion and cooked to give firm gels for slicing. IN starches can be used in cold-formed products, and CU in cooked recipes.

The use of starches in texturising chicken breast and similar portions of raw meat has been reported.

Starch handbook

Chapter 7

PERFORMANCE OF STARCH-BASED MATERIALS IN HARD AND BRITTLE FOODS

7.1 Overview

The creation of fluid batters and viscoelastic doughs from starch-rich raw materials that can be shaped into products is possible because of the solubilising and plasticising effects of water. This can occur over a range of temperatures after the crystalline structure of starch is melted, but water is necessary to allow the starch polymers to become mobile and to take on new forms to create structures in products.

After the structures have formed they can be transformed from viscous fluids to soft gels and on through rubbery, leathery and plastic states to become hard brittle glasses by drying to moisture contents of <5-10%. Removal of water stabilises the structure of a product by increasing its viscosity to an infinite level and then by reducing the volume available in the matrix for the movement of starch molecules. Large starch polymers eventually become immobile at moisture levels of <5-10% w/w, even at elevated temperatures (40-50°C), and on cooling to ambient temperatures, starch-rich structures become glassy and rigid (Chapter 1, 1.9). In the glassy state, they create a hard and brittle texture in the solid phase of products. They fracture with a sharp clean break and shatter into fragments like glass.

Snacks, breakfast cereals and breadings are all made with a low final moisture content to develop this hard brittle cell wall material, but because they have fine sponge cell structures, with thin cell walls, they are pleasant to bite. They are not too hard to break with the teeth and fracturing their cells may makes a crunchy noise during the first bite.

Biscuits have a denser structure than snack foods because most of their starch remains present as uncooked granules and their continuous glassy phase that controls the texture is based on sugar, or sugar and protein structures. They vary in texture according to the recipe and processing from being crumbly as short dough biscuits to the very hard bite of a gingersnap.

Confectionery products that are made using starches as texturising agents, such as jellies, gums, pastilles and liquorice, have high solids and sugar levels, and little or no aeration. They are dehydrated to about 85% solids, when they form a range of textures, from soft to hard rubbery gels. If they are dried to lower moistures as in capsules and coatings they become hard and glassy.

Low moisture foods can be classified in two groups:

1. Foods made from high or medium moisture doughs and batters that are cooked or baked, and then dried to their final moisture contents in the range 3-10% w/w. Examples of this type of product are biscuits, snacks, breakfast cereals, breading crumbs, croutons, melba toast, breadsticks, popodoms.

2. Foods that are made at a low moisture levels (0-20%) and are manufactured with a minimum of drying to achieve their final moisture content. These products might be either various forms of sugar confectionery, where there is a small amount of drying out, or tablets, where the moisture is always very low.

7.2 Biscuits

The large range of products made for the biscuits market either have recipes rich in fats and sugar, such as short dough, semi-sweet biscuits and cookies, or leaner recipes as in crackers, matzos, wafers and water biscuits (Manley, 1998).

7.2.1 Short dough biscuits and cookies

On a flour weight basis, short dough biscuit dough contains about 35% fat, 20% moisture and 25% sugar. This increases the gelatinisation temperature of the starch to close to or over 100°C. In cookies, the sugar level is higher at 50% of flour weight and therefore starch is not gelatinised in the dough during baking.

Table 7.1: Methods used to manufacture hard brittle products

Products		Baking cooking	Extrusion	Puffing	Frying	Other
Biscuits	Short dough and cookies	O				
	Semi-sweet	O				
	Crackers	O				
	Wafers					O
Breadings	Conventional baked	O				
	Japanese	O	O			O
	Cracker	O				
	Extruded		O			
	Dielectric/microwave					O
Breads special	Toasts	O				
	Breadsticks	O				
	Croutons	O			O	
Breakfast cereals	Flakes		#			O
	Puffed grain		#	O		
	Shredded		#			O
	Extruded		O			
Snacks	Crisps				O	
	Extruded direct		O			
	Extruded pellets		O			
	Fried dough				O	
	Puffed cereals			O		
	Rice cakes			O		
	Popodoms				O	
	Pretzels	O				
Confectionery	Pastilles					O
	Gums					O
	Liquorice		O			O

\# similar type of product made by extrusion

O manufacturing methods

In both these high fat and sugar recipes, the starch granules remain largely unchanged in the baked dough. Some are bound in the larger particles of flour with wedge protein and others are present as free granules. They retain most of their crystallinity and do not swell unless damaged by milling.

The high fat levels serve to hinder both the development of the proteins into a continuous phase and the break down of the flour particles. It is the sugar syrup that links the starch and flour particles together. After baking off the water to a low moisture level of 2-3%w/w, a sugar glass forms on cooling to give a brittle structure between the flour particles. In moister cookies the sugar recrystallises to give a hard texture during storage.

7.2.2 Semi-sweet

The lower fat content (16% on flour weight) in semi-sweet biscuits permits some hydration and development of the proteins to form a weak protein network. This is important for the sheeting of the dough into thin layers for the traditional sheeting/cutting process. The starch granules are embedded in these networks, but do not have strong contacts between each other. At the lower sugar levels (15%) for the same water level (20%) based on flour weight, as in short dough biscuits, there is a small amount of starch gelatinisation. This occurs mainly in the centre of the biscuits for up to 30-35% of the starch present (Burt and Fearn, 1983). The gelatinised starch is not able to swell in the biscuits, but can swell in the mouth as the biscuit is masticated to soften the texture.

7.2.3 Crackers

Cream crackers, which have simple bread-like recipes (enriched with fat to about 12% fw) with water levels of 30-40% of the dough, have a well-developed protein structure around their starch granules. The fermentation process allows the proteins to hydrate fully and some proteases are formed to soften the dough. During baking, most of the starch granules within the crackers are gelatinised, but remain embedded in gluten proteins, the exception being at the edges of the biscuits, where water is lost very quickly. The hard brittle structure of crackers is dominated by the glassy state of the gluten phase. It was reported that, for this type of product, the A_w should be <0.39 to give a crisp texture. However, there are differences in texture between the surface layers and the central region of crackers that has an important influence on eating characteristics. In penetrometry measurements with a cylindrical probe, it was shown (Guy and Anstis, 2001) that the force distance graph for a cracker reveals the main textural qualities. It shows a fast rise to peak force for the hard brittle outer layers, followed by slower decay of the force as it falls to zero and measures residual relaxation of the softer central layers. The tail of the graph represents the presence of a slightly chewy centre to the cracker, which is a desirable characteristic in the cream cracker.

7.2.4 Wafer biscuits

Wafers biscuits that are used in layered biscuits with chocolate or cream, or for ice cream cones, have a crisp texture that is easily shattered. They are formed from a batter of flour and water, with small amounts of sugar and fat, at about 150% water on flour weight. Batter slurry is prepared to a fixed viscosity, prior to depositing it on the lower plate of a pair of hot wafer plates or moulds. The plates have channels cut in their surfaces leading to their outer edges. After closing the plates, they are heated to 190°C for about 90s. Inside the plates, the water boils, gelatinises the starch and evaporates from the plates via the channels. The starch granules swell in excess water and are completely dispersed by the shearing forces developed in the paste as water boils and its vapour escapes through the channels to the atmosphere outside the plates. The dry starch left in the plates is completely amorphous with no traces of any starch granules and forms an alveolar structure. At low moisture, <5%, the texture is crisp and brittle because the amorphous starch polymers are in the glassy state, but as moisture increases above 8-10% the texture becomes plastic and leathery. The structure is very porous and readily takes up moisture from the atmosphere to lose its crispness.

7.3 Breakfast cereals

Breakfast cereals (Fast, 2000) are formed from moist grains or doughs by a number of ingenious processes, including flaking, puffing, hot water cooking, shredding and extrusion cooking.

The main types in the market are

- Flakes: made from maize, wheat bran, wheat or mixed grains.
- Puffed cereals: made from wheat and rice grains and specially prepared mixed flours doughs.
- Extruded puffs and flakes: made from maize, wheat, rice and mixtures of many cereal and starch types.

7.3.1 Flaked products

Corn flakes, one of the original breakfast cereals from *circa* 1900, are made by heating large hominy maize grits, with about 30% of their weight of water and 6-8% of small sugars, under pressure in a sealed rotary cooker for 2-3h at 101-108°C (1-1.25 bar). The starch granules are all gelatinised but do not swell very much; the grits become soft and after conditioning at ambient, or lower temperatures, and can be flaked to thin sheets under large flaking rolls. The flakes are toasted to remove water and leave a dry brittle structure with gelatinised granules embedded in the cellular structure of the plant.

Wheat products require less cooking time in the rotary cooker (about 30min) because the penetration of hot water is easier through their more porous endosperm. In mixed systems, some care must be taken with combinations of cereal types used to ensure that the harder and more vitreous endosperm types are fully cooked, without breakdown of the softer types due to overcooking and the formation of sticky lumps in the cooker.

The flakes may be formed into large biscuits immediately after flaking, while they are still hot. The small flakes are packed into rotary moulds (using biscuit technology), or compressed under metal belts and cut to form moist biscuits,

2-3cm thick. These are dried to reduce the moisture down to <5% to form a brittle crunchy product.

7.3.2 Puffed cereals

These products are made from whole grains, such as rice (Rice Krispies), durum wheat or bread wheat (Puffed Wheat) by cooking the materials at low moisture 16-20% in a sealed chamber at high temperature, 180-190°C (up to 14 bar). The starch granules within the grains lose their crystallinity and become amorphous in the first stage of gelatinisation, but do not swell due to lack of water. On the firing of the gun (opening of the pressure vessel), the water evaporates out of the grain using the holes in the starch granules as nuclei for bubble expansion. In the expanded product many of the granules are dispersed entirely to form individual gas cells, which are very numerous in the finely textured sponge structure. Sufficient moisture is lost to form a stiff texture on cooling at 5-7% moisture w/w, but further drying is used to reduce the moisture to about 3% w/w.

Some puffed products are expanded in the same manner from special dough pellets. These are manufactured on an extrusion cooker using mixtures of cereals at about 20% w/w moisture, with sufficient heat to melt all the crystallinity in the starch granules. These pellets can be expanded in a puffing gun or chamber in the same way as the grain. It is important that the extrusion cooking does not destroy the granular structure in the dough, so that the pores remain in the granules (see Chapter 1, 1.6) as nuclei for expansion.

7.3.3 Hot water products

One of the best known of the original breakfast cereals (*circa*, 1894) is shredded grain (Shredded Wheat). It is made by cooking wheat grains in hot water at atmospheric pressure, until the starch granules at the centres of the grains are gelatinised and the grain moisture rises to 45-50% w/w. The softened grains are drained and allowed to stand for 24h to allow for the redistribution of moisture. They also firm up a little, because some retrogradation of the starch takes place, and at that stage they are ready for the shredding process. The soft kernels are rolled between a smooth roll and a grooved roll and the shreds are removed with a combing device to produce a thin sheet of shredded wheat. A large biscuit can be built up by overlaying layers from 10-20 pairs of rolls and cutting the web into the required sizes. Smaller bite sized products require fewer layers to achieve the products. The moist biscuits are dried to <3% moisture in large ovens and become crisp and brittle.

Extruded cereals

The extrusion cooker is a processing unit that can transform cereal mixes into a molten fluid dough at 18-25% moisture. The molten fluid can be used to manufacture a range of products similar in nature to those made by older conventional cooking processes. There are two main forms of the extrusion process.

- Directly extruded products are formed as balls, hoops and more complex shapes by cooking cereal mixes at medium shear and high temperatures and expanding the cooked dough directly at the die. The same process is used to make single and multigrain products and others with variations in nutrition and texture.

- Pellet extrusion is a two-stage process that begins with the formation of dense dough at low shear and high temperatures. The dough is extruded at <100°C so that it does not expand and is cut into small beads or pellets. These are used in secondary processing to form strands for shredded cereal products. They may also be flaked on rolls to produce cornflakes, or any other single or mixed cereal flake.

Direct expansion

The feedstock for the extruder may be formulated from flours and grits and blended from several cereal types with added fibres and proteins. The processing moisture is a key variable for controlling both the temperature for gelatinisation and the frictional/shearing forces within the extruder. Normally, a moisture level of 22-25% w/w on the dough is used to create conditions in the extruder of about 150°C and pressures of 10-15 bar. The mechanical energy input is important for the control of the partial dispersal of starch granules to form a continuous starch phase. This should be in the range 400-500kJ/g specific mechanical energy to give enough dispersal of starch for expansion but to retain 50-60% of the granular structure for good eating quality. RVA profiles can be used to determine the level of cook in the starch and help to adjust the process to the optimum state. The temperature is also important and should be in the range 140-160°C to achieve good flavour development (Bredie *et al.* 1998a; b).

The developed fluid can be expanded at the die, if it is extruded at >105°C (preferably 140°C), to give a specific volume of 3-4ml/g and the extrudate can be dried to <3% moisture to give the finished product. Good control of the extrusion process is required to form expanded sponge structures with fairly thick cell walls to resist softening and retain crunchiness when soaked in milk. Ideally, the products should retain most of their hard crisp structures for a few minutes while being consumed.

Pellet processing

The development of the starch for pellet doughs is carried out at 110-120°C and pressures of 5-10 bar, with moisture levels of 26-30%, w/w. For good product quality, all the starch granules in the mix should be melted to lose their crystallinity, but they should only be degraded a little from their granular form. This requires lower specific mechanical energy (SME) inputs than for direct expansion, in the range 250-300kJ/kg. The RVA graphs for the pellet extrusion should show a high cold paste viscosity, retaining a small peak in the graph. They should match the profiles obtained when cereals are cooked in rotary pressure vessels, as in the traditional processes.

If the cooked dough is extruded at <100°C, it will not expand at all, but can be cut into dense beads. These remain soft and plastic at 50-60°C and can be fed into shredding rolls to form sheets of shreds as in the conventional process. The sheets can be combined to form a lattice or shredded biscuit product and toasted in an impingement dryer to <3% moisture to form dry crunchy products. Pellets can also be flaked at about 60°C to form cornflakes and other types of flakes made from mixtures of cereals, or enriched with bran and other fibres or proteins. The dimensions can be controlled by the die diameter used to form the pellets.

7.4 Breadings

7.4.1 Conventional bread technology (home style)

Breadcrumbs are used in cookery for coating the surfaces of fried products and as water binders in several types of meat products, such as sausages and burgers. Commercial breadings for the food industry are made from yeast-raised doughs in roughly the same types of process as used for conventional bread. However, the loaves are made to different dimensions and generally are denser than normal standard bread. The baked bread has a crumb structure with a continuous gluten phase surrounding gelatinised starch granules as in normal bread. The soft crumb is shredded with rotating blades and dried to low moisture (<10% w/w) as crumbs which can be sieved into the required size ranges. The character of the crumbs can be varied by manipulating the baking process to produce light coating crumb and denser stuffing crumb for water binding in meat pastes such as sausages.

7.4.2 Conventional bread technology (Japanese Panko)

A special light textured crumb was invented in Japan for their traditional products. Originally it was made by over-proving their doughs and baking in tall tins. This gave long thin cells that could be shredded to give breadcrumbs with needle-shaped fragments, known as Panko or Japanese crumb. The use of special baking technologies has improved the production of this type of crumb. If the dough is proved in special cells with a pair of metal sides it can be baked by dielectric heating for 10min to give a white crustless loaf. The crumb can also be made from sheets of unproven dough using microwave heating to give instant lift to the dough instead of the conventional yeast-raised technology.

7.4.3 Cracker technology

Simple doughs of flour, salt, reducing sugar, colour and water (35%) are prepared in a continuous mixer and sheeted down to 25mm before baking. The product is crumbled on a granulator and dried to about 8% moisture and sieved to obtain the required size ranges. It has a proportion of gelatinised starch, which can absorb water, and this factor can be varied to give a range of water absorptions. It is used mainly for water binding in meat products.

7.4.4 Extrusion cooked breading

An extruded product can be made that is similar to normal breadcrumb. A sponge structure with a cell structure similar to baked bread can be formed by extrusion cooking flour at 30% moisture content, w/w, at 120-130°C with a low mechanical energy input of 100-150kJ/kg. The process is more efficient than breadmaking at 45% moisture, w/w, and more flexible because it is possible to use many different types of starch-rich materials, or mixtures of such materials, to form the crumb. For example, products can be made from wheat, maize or rice, or combinations of all these materials, or other starch-rich materials such as potato and cassava. The extrudate is shredded while moist and dried to <10% moisture, when it becomes brittle and glassy.

However, the crumb structure is completely different from conventional breadcrumb, being formed from disrupted starch granules surrounding cooked intact granules and dispersed gluten (Guy, 1986). There is no continuous protein phase, as found in conventional crumb, but the specific volume can be adjusted to similar values to conventional bread at 3-4ml/g.

The difference in the nature of the continuous phases in extruded and conventional breadcrumb may introduce some problems in the extruded crumb not found in breadcrumb. These are evident when the crumb is fried; the appearance is dull and the tips become white, often described as "frosting". The dullness can be related to the presence of tiny air bubbles entrapped in the cell walls of the crumb. Starch cell walls are clear like glass unless these bubbles are present, due largely to extruding at too low a moisture content, <26% w/w.

Frosting is a remarkable change in the starch glass, which CCFRA associates with changes in moisture in the tips of the crumb during frying. If the moisture in the extruded crumb is 8-10% w/w, frosting may occur, and if moisture is >10%, it is almost certain to occur.

7.5 Breadsticks, croutons and toast

Breadsticks, croutons and toasts are made in a similar manner by preparing dough by conventional mixing and proving technologies, with the additional step of oven drying to produce a crisp texture and stable product.

7.5.1 Breadsticks or Grisini

These products are made by a special moulding process to form the sticks. Proved bread dough is sheeted out between pairs of parallel grooved rollers to form individual smooth surfaced rods of dough. These are spread apart 2-3cm by a special wire belt to allow further expansion and permit full dehydration in the oven. After cutting to length, the sticks are deposited on trays and baked and dried out in an oven to <10% moisture.

Alternative methods with extruded doughs can be used but they tend to form denser products because the pressure on the dough removes some of the gas bubble structure.

7.5.2 Croutons

The crouton is manufactured from bread that is formed from yeast-raised dough and baked to form a crumb structure. The crumb is staled for up to 36h, cut into small cubes and dried to <5% moisture, w/w, in a forced convection oven at 190°C. There are several forms of crouton but they may be divided into two types, for applications either in soups or salad.

Soup croutons are pieces of breadcrumb that are fried at 130°C in fairly hard fats. This creates a protective layer at their surfaces to extend their life in hot soup for a few minutes before the texture softens. The fat content of the traditional product is high, at 40%, but lower fat contents are becoming available.

Salad croutons are formed from bread that is coated with 10-15% oil and baked, or toasted, to low moisture to give a crunchy texture and colour. These products are made in several sizes and shapes from cubes to shredded crumbs.

7.5.3 Toasts manufactured from bread

Toasts are prepared as loaves of bread of the required cross-sectional shape and dimensions. After baking, a bread-like crumb structure is formed, which is the basis of the toast. It may be allowed to stale for a set time (10-24h at 4°C) to firm the texture, before slicing and toasting. This is more important for bread with a high specific volume made to give a light textured toast. The texture of toasts is similar to bread, with a continuous film of gluten surrounding gelatinised starch granules.

7.6 Snack foods

7.6.1 Crisps and chips

Crisps are prepared from fresh cut potatoes, either by washing the cut slices, or by cutting and projecting them directly into the hot oil. The former process removes loose starch and soluble sugars to reduce materials in the oil and browning in the crisp. After frying through a cycle finishing at 170-180°C, the crisps have a low moisture content of 1.5-2% w/w and have a glassy and brittle texture.

Aguilera *et al.* (2001) heated potato cells in oil on a hot stage microscope equipped for video-microscopy. They found that starch granules in the plant cells underwent rapid gelatinisation, deformation and compaction into a dense mass that occupied the whole volume of each cell. There were no signs of any disruption of cell walls, but at temperatures >100°C, a reduction in cell area occurred. After dehydration, cells showed a distinctive outer zone and a homogeneous core.

Some other raw fruits and vegetables are also used to manufacture crisps, including apple, sweet potato and carrots, but potato is the by far the most important type.

7.6.2 Extruded snacks made by direct extrusion

These are similar to extruded breading crumb and breakfast cereals in that they are expanded from doughs developed in an extrusion cooker. However, they are made at lower moisture levels, 15-18% w/w, than breakfast cereals and have specific volumes of up to 6-7 ml/g. Normally, they are made from basic starch-rich materials such as maize grits, potato flakes/grits, wheat, cassava or rice flour. Microscopic and gel permeation chromatography (GPC) examination of the starch polymer structures in the products revealed that more than half of the starch granules are completely dispersed and their amylopectin polymers are broken down to about 1MDa (Guy *et al.*, 1996). This form of dough gives highly expandable cell wall materials and allows large numbers of cells to be nucleated to produce finely structured alveolar cell structures.

The choice of a single raw material, such as flour, will require information about its composition and physical performance in the process. Recent studies with maize, rice and wheat flours at CCFRA have shown that the hardness of the endosperm is important in determining the severity of the processing conditions within an extruder. Major cereals, such as wheat, maize and rice, exhibit variations in performance due to natural variations in the endosperm texture or starch composition. Maize grits tend to have a fairly consistent hard endosperm texture, which gives small variations in frictional heating and shearing effects in an extruder (Guy, 2001). However, the composition of starch may vary widely in its proportions of amylose and amylopectin, giving significant variations in expansion and texture formation. Rice flours and grits milled from native grains tend to display similar compositional variations as maize.

Wheat flours show little variation in starch composition, but have a wide range of variation in endosperm texture. Flours from soft wheats create less heat and shear in the compression zone of an extruder than those from hard wheats. This results in different process conditions for the same machine settings and a change in the amount of dispersal of the starch polymers from the granules. Therefore, the wheat flour should be chosen as a blend of hard and soft wheats that suits the extruder and gives the required structure.

Directly extruded snacks can be made from any basic starch-rich material with few problems, but if the texture needs modification it is also possible to add some special materials. The tuber-based materials have very low lipid contents and always need some added oil or emulsifier to ensure smooth extrusion over the hot metal surfaces.

High amylose starches can increase expansion and bubbles in the texture. Some potato derivatives with weak granular structures can also increase expansion if the extruder is not very flexible.

Little drying is required for directly extruded products as the residual moisture levels are usually <8-10% w/w. The cooled starch sponge structures are brittle and crispy/crunchy. They are usually dried to <5% moisture w/w to ensure a long life.

7.6.3 Extruded by pellet methods

The pellet, or half product, process is performed in three stages. Firstly, a starch and water dough is cooked on an extruder to melt the crystalline structure of any starches present and the dough is extruded at 90-95°C without expansion. Secondly, the extrudate is cut into pellets and these are dried to about 10% moisture to form the half products. The extrudate may be cut at the die face, or formed into a sheet that is cut after further treatment with steam. Finally, the half product is expanded by frying in oil (or heating in a fluidised bed hot air expander) at 170-190°C for 10-15s to form the finished snack. The snack pellet processes can be performed with basic raw materials, such as maize, rice, cassava, potato and wheat flours. In the full cooking process at about 25-30%

moisture w/w, the native flours or blends are cooked to melt the crystallinity within the starch granules, but not to disperse the granules. Some special starches are available from the specialist starch manufacturers to improve the performance of pellets. These may help expansion and texture control.

The process can also be performed with pregelled raw materials, such as combinations of potato granules and flake, in a low temperature extrusion or sheeting process at similar moisture levels.

7.6.4 Fried snacks formed from moist dough

The manufacture of snacks such as tortilla chips, chipsticks and stacking crisps is based on the formation of a dough from precooked maize or potato derivatives. The dough may be mixed in a simple mixer or an extruder, and individual snack pieces formed from it by extrusion, or sheeting and cutting. Moist snack pieces, which may be dried in some cases, are fried in oil at 170-190°C to flash off the moisture and create a crisp brittle texture. Drying the snacks before frying significantly reduces the retention of oil (Moreira *et al.* 1997)

The basic raw materials for the maize products are masa flours developed by cooking whole maize, or by a variation using maize medium/fine grits (Bello-Perez *et al.* 2003). Small amounts of other starches can be used at low levels to enhance the binding in the dough to improve sheeting performance and the puffing of the snacks in the fryer. Masa flours are ground to a coarse particle size and mixed with finer fractions to obtain a good texture in the fried products. The use of more soluble starch additives, such as pregelled maize flour, helps to bind the larger particles together at the dough stage.

Potato-based products have similar qualities and form the basic raw materials for tubes, rings, sticks and stackers. The doughs are formed by adding water to a combination of potato granules, potato flake and modified potato starches, and mixing vigorously to cause shear damage and induce stickiness. The large coarse potato granules give a good eating quality but are not cohesive enough to form a dough and need to be stuck together with the cohesive flake. Pregelled forms of potato starch are sometimes used to improve binding quality and to give better expansion on frying.

7.6.5 Puffed snacks

The puffing of starch granules occurs in simple products such as popcorn and puffed wheat. Other products have been developed from doughs dried to form glassy structures (as mentioned in 7.5.2).

All these products share a common method of structure formation despite differences in their form. The main principle for puffing is to heat a pellet of starch-based material that is in the glassy state at low moisture (10-16%w/w), until the temperature of the glassy structure is surpassed. This will be in the range 140-170°C and the water within the structure will be a super heated liquid (7-8 bar at over pressure). In the popping corn, the shell of the grain controls the over pressure, but for other products a chamber is used to prevent the water flashing off as vapour before the starch has melted and softened the endosperm texture. Opening the chamber after the melting temperature has been reached causes the water to vaporise and expand the starch granules from the holes at their hilia.

7.6.6 Puffed rice cakes

A special form of expanded snack has been developed from rice grains that are closely related to popcorn and puffed wheat. Rice grains, or kibbled grains, are sealed in a shallow cylindrical mould consisting of a ring to form the sides and moveable top and bottom platens. After the mould is heated to about 200°C (Hsieh *et al*. 1989), the upper platen is moved rapidly to the upper edge of the ring, releasing the pressure. The rice grains with soft melted starch granules expand explosively to fill the space and form a flat disc of "rice cake". The individual rice grains expand and press into each other in the limited space available, to lock their structures together and form a crisp hard cake on cooling. In production, a large number of moulds must be used to give good output.

Smaller versions of the original rice cakes have been developed as snacks and flavoured with both sweet and savoury coatings. The expanded texture contains many very small cells probably formed from individual rice starch granules.

7.6.7 Popodoms (papads)

This traditional Indian snack is manufactured in several forms, mostly as discs with sizes from 5-25cm diameter. The smaller forms have been adopted for the snack industry while the larger sizes are used in the home and in the restaurant trade.

The main ingredients are gram flours, usually black gram flour, salt and water, but some products are also made with 25% rice flour, or other gram flour types. The water is mixed with the flours to form dough at 40-45% fw, with a minimum of mixing. The dough is rested for several hours and sheeted out to 0.5mm and dried at 50-60°C until the moisture is between 12 and 14%. The papad is stable for many months at this moisture and may be fried at 180-190°C to expand by 50% in diameter and thickness. The papads may also be expanded in a microwave oven.

7.6.8 Hard Pretzels

Pretzels are manufactured from soft wheat flour by mixing with water (42% fw) and yeast to form a gluten network and resting for 20min and extruding to form the classic looped shapes or straight sticks. The formed dough pieces are immersed in a bath of 1% sodium hydroxide and heated at 88-93°C for 30-60s. After leaving the caustic bath they are sprinkled with nubs of salt, before baking at high temperatures (280°C) for 4-7min. The moisture is reduced to 10-12% and a glossy brown colour develops at the crust, due to the solubilisation of surface starch. A fast Maillard reaction at the high pH creates colour and reduces the pH to 7.0. The final drying at 110-112°C reduces the moisture to <4%. The only gelatinised starch in a pretzel should be in the surface layer because the water is dried off too quickly for the starch in the interior to gelatinise (Seetharaman *et al.* 2002).

7.7 Products made from low moisture doughs - confectionery

Starch is used in confectionery in many forms such as glucose syrups, maltodextrins, acid-thinned, oxidised- and native starch. The role of starch is mainly to change the texture of sugar syrups to become more viscous and even develop a gelled state. The large amounts of sugars present move T_g for the system to much lower levels, so that for many products the physical form of the starch polymers is a rubbery state above the glass transition. Gelling takes place between the smaller almost linear chains of amylose and the degraded amylopectin. The main product types where starch contributes to texture are the "Hard gums", softer "Jelly pastilles" and liquorice products.

7.7.1 Gelled gum products

The ability of starch to form gels depends on its molecular composition and modification by processing. It has been shown that amylose can form strong gels at high concentrations, whereas amylopectin, although creating large viscosity effects in water, does not form gels at a rate that is useful in confectionery. Several forms of confectionery utilise the gelling properties of amylose, or the degraded forms of amylopectin which are more linear and more like amylose in character.

The commercial products used in the industry were originally based on acid-converted maize starch of 60-70% fluidity (3B.2). In these materials, the starch polymers are degraded so that their cooked viscosity is low and high levels of these converted starches can be added to recipes, in the range 10-15% w/w. These mixtures can be processed at 95-105°C and deposited at high solids (82-84% w/w).

High amylose maize starches, with amylose contents of 50-70%, have low viscosities and strong gelling power in the hot confectionery slurries. They can be used as a minor component of the starch blend to give significant benefits, in significantly reducing gelling time and improving gel strength. However, they require a higher processing temperature (168°C) in the cook to gelatinise the granules because of the high solids levels. Therefore, further improvements have been made to reduce the cooking temperature and viscosity while keeping the faster setting action.

The gum products may be divided into two types: Jelly gums and Hard gums. Typical levels of starch in Jelly gums are in the range 10-15%, but for Hard gums the levels are increased to 20-30%. This means that for the Hard gums the basic viscosity of the starch must be much lower and so low-fluidity starch is used. The amount of degradation of the starch polymers is increased in the low fluidity dextrins by more severe treatments with either acids or enzymes.

7.7.2 Coatings and glazes

Coatings and glazes were originally based on a mixture of hydrocolloid polymers, sugar solids and small dextrins, such as maltodextrins. The hydrocolloids, such as gum acacia or cellulose ethers, formed the main structures of the films in the coatings as the water was removed. Amylose has been shown to be a good film-forming polymer and the original high amylose maizes were developed for this purpose. However, pyrodextrins which have a low viscosity, but good film forming ability, have been used for many years in this role and can be fortified with high amylose starches.

7.7.3 Marshmallows

Marshmallows are gelatin stabilised sugar foams, made from sugar and glucose syrup. Addition of starch stiffens and dries out the protein cell walls to stabilise the foam. In the process the starch, water and sugar are heated to about 118°C to gelatinise the starch. The syrup is added and the temperature reduced to about 88°C, before adding gelatin as solution. The mixture is aerated at about 42-43°C at a pressure of 2 bar in a scraped surface beater/heat exchanger. Finally, the marshmallow is extruded into a starch bed, and sets quickly when starch is present. The best results are obtained with a cross-linked waxy maize starch

7.7.4 Liquorice

Liquorice is based on wheat flour, liquorice root, sugar, sugar syrups, molasses and water. In an open pan method, the mixture is processed with 40% w/w water and 20% w/w wheat flour to gelatinise the starch. The dough is cooked for 2-3h until the solids level rises to about 82%. Extrusion cookers with their ability to process highly viscous doughs can be used to manufacture liquorice from doughs containing only 20% moisture to form the product in 40-60s in a continuous stream. The main process requires a high temperature of >140°C to melt the crystallites in the starch granules but the shear inputs must be controlled to avoid over shearing the dough. A particular feature of good liquorice texture is the presence of gelatinised granules in the finished product. Undercooking or over shearing reduces the eating quality. Speciality cross-linked starches have been recommended at 5-7% of the flour weight to reinforce the starch population and help retain the population of gelatinised granules.

These are typical confectionery processes, where starch is used at 30-40% with sugar/glucose at 60-80% of solids to give a high viscosity in a recipe. The moisture must be low at 20-22% w/w. Special starches that have been degraded or converted as in 3.2.1 give more control in the process and reduce the energy requirements of the extruder.

Chapter 8

REFERENCE LIST

Abdd Karim, A., Norziah, M.H. and Seow, C.C. (2000). Methods for the study of starch retrogradation. *Food Chemistry*, **71**, 9-36.

Agboola, S.O., Akingbala, J.O., and Oguntimien, G.B. (1997). Physicochemical and functional properties of low DS cassava starch acetates and citrates. *Starch,* **43**, 62-66.

Aguilera, J.M., Cadoche, L., Lopez, C., and Gutierrez, G. (2001). Microstructural changes of potato cells and starch granules heated in oil. *Food Research International*, **34**, 939-947.

Anderson, C., Catterall, P.F. and Clarke,C.I. (2000). Heat-treated cake flour club final report. BCP/44679/1, CCFRA, Chipping Campden. pps1-60

Anderson, R.A., Conway, H.F., Pfeifer, V.F. and Griffin, E.J. (1969). Gelatinisation of corn grits by roll and extrusion cooking. *Cereal Science Today*, **14**, 4-7; 11-12

Badenhuizen, N.P. (1965). Occurrence and development of starch in plants. In 'Starch Chemistry and Technology Volume 1: Fundamental Aspects', R.L. Whistler and E.F.Paschall, Eds. Academic Press, New York, pp. 65-66.

Bauer, B.A., Wiehle, T. and Knorr, D. (2005). Impact of high hydrostatic pressure treatment on the resistant starch content of wheat. *Starch,* **57**, 124-133.

Bean, M.M., Donelson, D.H. and Yamazaki, W.T. (1978). Wheat starch gelatinisation in sugar solution II, fructose, glucose and sucrose: Cake performance. *Cereal Chemistry*, **55**, 945-952.

Bello-Perez, L.A., Osorio-Diaz, P., Agama-Acevedo, E., Solorza-Feira, J., Toro-Vazquez, J.F. and Paredes-Lopez, O (2003). Chemical and phyicochemical properties of dried wet masa and dry masa flour. *Journal of the Science of Food and Agriculture*, **83** (5), 408-412

Bhatti, M. and Hall, A. (2002). Development of standard tests for the texture measurement of baked products. CCFRA Research Report 159, 1-52.

Bradbury, A.G.W. and Bello, A.B. (1993). Determination of molecular size distribution of starch and debranched starch by a single procedure using high performance size exclusion chromatography. *Cereal Chemistry,* **70**, 543-547.

Bredie, W.L.P., Mottram, D.S. and Guy, R.C.E. (1998a). Aroma volatiles generated during extrusion cooking of maize flour. *Journal of Agricultural and Food Chemistry*, **46**, 1479-1487.

Bredie, W.L.P., Mottram, D.S., Hassell, G.M. and Guy, R.C.E. (1998b). Sensory characterisation of the aromas generated in extruded maize and wheat flour. *Journal of Cereal Science*, **28**, 97-106.

Brock, C. and Greenwell P. (1994). Acetylation of wheat flour to improve cake flour performance in high-ratio cakes. *FMBRA Final Report* to the Consortium.studying methods for improving high-ratio cake flour.

Buhler (2005). Gluten-free pasta. *Food Engineering and Ingredients*, **30** (6) 19.

Burt, D. J. and Fearn, T. (1983). A quantitative study of biscuit microstructure. *FMBRA Bulletin,* (1) 31-38.

Cauvain, S. P., Cyster, J. A. and Hodge, D. G. (1979). The heat treatment of wheat or semolina as an alternative to chlorination. *FMBRA Research Report* **83**, 1-31.

Chamberlain, N. (1962). The behaviour of starch suspensions as an explanation of cake structure. *BBIRA Bulletin,* (6) 164-165.

Chiu, C. W., Schiermeyer, E., Thomas, D. J. and Shah, M. B. (1998). Thermally inhibited starches and flours and process for their production. *US Patent* 5,725,676.

Conde-Petit, B., Nuessli, J., Handschin, S. and Escher, F. (1998). Comparative characterisation of aqueous starch dispersions by light microscopy, rheometry and iodine binding behaviour. *Starch,* **50**, 184-192.

Delcour, J. A.(2002). Recent advance in enzymes. In 'Grain Processing', Courtin, C. M., Veraverbeke, W. S. and Delcour, J. A., Eds. KUL, Leuven, Belgium.

Donald, A.M. (2005). Understanding starch structure and functionality. In 'Starch in Food: Structure, Function and Applications', A.C.Eliasson, Ed. Woodhead Publishing Ltd, Cambridge, pp. 156-184.

Douzals, J.P., Marechal, P.A., Coquille, J.C. and Gervais, P. (1996). Microscopic study of starch gelatinisation under high hydrostatic pressure. *Journal of Agricultural and Food Chemistry,* **44**, 1403-1408.

References

Dull, B.J. (2001). Natural, non-GMO baking ingredient from rice bran. *Zucker und Suesswaren Wirtschaft,* **54**, 16-17.

Fannon, J.E., Gray, J.A., Gunawan, N., Huber, K.C. and BeMiller, J.N. (2003). The channels of starch granules. *Food Science and Biotechnology,* **12**, 700-704.

Fannon, J.E., Hauber, R.J. and BeMiller, J.N. (1982). Surface pores of starch granules. *Cereal Chemistry,* **69**, 284-288.

Fast, R.F. (2000). Manufacturing technology of ready-to-eat cereals. In 'Breakfast Cereals and How They Are Made', R.F. Fast and E.F.Caldwell, Eds. AACC, St. Paul, Minn, pp. 17-54.

Fitt, L.E. and Snyder, E.M. (1984). Photographs of starches. In 'Starch Chemistry and Technology', R.L. Whistler, J.N. BeMiller and E.F. Paschall, Eds. Academic Press, New York, pp. 675-689.

Frazier, P.J., Brimblecombe, F.A. and Daniels, N.W.R. (1974). Rheological testing of high-ratio cake flour. *Chemistry & Industry (London)* 1008-1009.

Gianbelli, M. C., Sissons, M. J. and Batey, I. L. (2005). Effect of source and proportion of waxy starches on pasta cooking quality. *Cereal Chemistry,* **82** (3), 321-327.

Gomes, M.R.A., Clark, R. and Ledward, D.A. (1998). Effects of high pressure on amylases and starch in wheat and barley flours. *Food Chemistry,* **63**, 363-372.

Graybosch, R. A., Souza, E., Berzonsky, W., Baenziger, P. S. and Chung, O. (2005). Functional properties of waxy wheat flours: genotypic and environmental effects. *Journal of Cereal Science,* **38** (July) 69-76.

Greenwell P. (1994). Starch granule functionality. The role of starch granule protein. *Chorleywood Digest* **135** (March) 40.

Guy, R.C.E. (1983). Factors affecting the staling of Madeira cake. *Journal of the Science of Food and Agriculture,* **44**, 474-491.

Guy, R.C.E. (1986). Extrusion cooking versus conventional baking. In 'Chemistry and Physics of Baking', J.M.V. Blanshard, P.J. Frazier, and T. Galliard, Eds. (London: The Royal Chemical Society), pp. 227-235.

Guy, R.C.E. (1995). Cereal processing: The baking of bread, cakes and pastries, and pasta production. In 'Physico-chemical Aspects of Food Processing', S.T.Beckett, Ed. (Glasgow: Blackie Academic & Professional), pp. 258-274.

Guy, R.C.E. (2001). Raw materials for extrusion cooking. In 'Extrusion Cooking: Technologies and Applications', R.C.E.Guy, Ed. (Cambridge: Woodhead Publishing), pp. 5-27.

Guy, R. C. E. and Anstis, J. A. (2001). Effects of enzymes in baking: applications to cakes and crackers. CCFRA Research Report **146**, 1-57.

Guy, R. C. E., Hodge, D. G. and Robb, J. (1983). An examination of the phenomena associated with cake staling. FMBRA Research Report **107** (Nov) 1-43.

Guy, R. C. E and Mair C. (1994). The improvement of wheat flour for use in high-ratio cakes - heat-moisture treatments. *FMBRA Final Report* to the Consortium.

Guy, R.C.E., Osborne, B.G., and Robert, P. (1996). The application of near infrared reflectance spectroscopy to measure the degree of processing in extrusion cooking processes. *Journal of Food Engineering,* **27**, 241-258.

Guy, R.C.E. and Pithawala, H.R. (1981). Rheological studies on the mechanism of cake flour improvement. *Journal of Food Technology,* **16**, 153-166.

Guy, R. C. E. and Sahi, S. S. (2005). Effects of manufacturing processes on the performance of starch: Milled cereal derivatives. CCFRA Research Report **224**, 1-48.

Guy, R. C. E and Sahi, S. S. (2006). Effects of manufacturing processes on the performance of starch: Heat-treated flours. to be published.

Hibi, Y., Matsumoto, T. and Hagiwara, S. (1993). Effect of high pressure on the crystalline structure of various starch granules. *Cereal Chemistry,* **70**, 671-677.

Hizukuri, S. (1986). Polymodal distribution of the chain lengths of amylopectin, and its significance. *Carbohydrate Research,* **147**, 342-347.

Hizukuri, S. and Maehara, Y. (1990). Fine structure of wheat amylopectin: the mode of A to B chain binding. *Carbohydrate Research,* **206**, 145-159.

Hogan, J.T. (1977). Rice and rice products. In 'Elements of Food Technology', N.W.Desrosier, Ed. AVI, Westport, CT.

Hsieh, F., Huff, H. E., Peng, I. C. and Marek, S. W. (1989). Puffing of rice cakes as influenced by tempering and heating conditions. *Journal of Food Science,* **54** (5) 1310-1312.

Huber, K.C. and BeMiller, J.N. (2001). Location of sites of reaction within starch granules. *Cereal Chemistry,* **78**, 173-180.

References

Jane, J-L., Kasemuwan, T., Leas, S., Zoebel, H. and Robyt, J. (2006). Anthology of starch granule morphology by scanning electron microscopy. *Starch*, **46**, 121-129.

Jane, J., Xu, A., Radosavlejvic, M. and Seib, P. (1992). Location of amylose in normal starch granules. I. Susceptibility of amylose and amylopectin to cross-linking reagents. *Cereal Chemistry*, **69**, 404-409.

Jayakody, L. and Hoover, R. (2002). The effect of lintnerization on cereal starch granules. *Food Research International*, **35**, 665-680.

Katz, J.R. (1930). X-ray studies on starch. *Zeitschrift für Physikalische Chemie*, **150**, 90.

Kovacs, E.T. (2005). The use of the enzyme transglutaminase for developing pasta products with high quality. In 'Using Cereal Science and Technology for the Benefit of Consumers - Proceedings of the 12th International ICC Cereal and Bread Congress, 23-26 May 2004, Harrogate, UK'. Cauvain, S.P., Salmon, S.S. and Young, L.S. Woodhead, Eds. Cambridge, UK pp498-503.

Krog, N. (1971). Amylose complexes with food grade emulsifiers. *Starch*, **23**, 206-210.

Kruger, J.E. (1996). Noodle quality - what can we learn from the chemistry of bread making? In 'Pasta and Noodle Technology', J.E.Kruger, R.B.Matsuo and J.W.Dick, Eds. AACC, St. Paul, Minn, pp. 157-168.

Levine, H. and Slade, L. (1988). Collapse phenomena - a unifying concept for interpreting the behaviour of low moisture foods. In 'Food Structure - Its Creation and Evaluation'. J.M.V. Blanshard and J.R. Mitchell, Eds. Butterworths, London, pp. 149-180.

Luh, B.S., Barber, S. and Benedito de Barber, C. (1991). Rice bran: chemistry and technology. In 'Rice Volume II: Utilization', B.S. Luh, Ed. Van Nostrand, New York, pp. 313-362.

Luh, B.S. and Mickus, R.R. (1991). Parboiled rice. In 'Rice Volume II: Utilization', B.S. Luh, Ed. Van Nostrand, New York, pp. 51-88.

Luh, B.S. (1991). Rice flours in baking. In 'Rice: Volume II: Utilization', B.S. Luh, Ed. Van Nostrand, New York, pp. 9-34.

Manley, D. (1998). Technology of Biscuits, Crackers and Cookies. D.Manley, Ed. Woodhead, Cambridge, pp. 1-499.

McCarthy, D.F., Gallagher, E., Gormley, T.R., Schober, T.J., and Arendt, E.K. (2005). Application of response surface methology in the development of gluten-free bread. *Cereal Chemistry*, **82**, 609-615.

Milatovic, L. and Mondelli, G. (1991). The role of the new hydrothermic processes (HT-THT) in pasta production. In Pasta Technology Today, Chirrotti Editori, Pinerolo, Italy, pp. 121-174.

Miles, M.J., Morris,V.J. and Ring, S.G. (1984). Some observations on the retrogradation of amylose. *Carbohydrate Polymers,* **4**, 73-77.

Moreira, R., Xiuzhi Sun, and Youhong Chen (1997). Factors affecting oil uptake in tortilla chips in deep-fat frying. *Journal of Food Engineering,* **31**, 485-498.

Morrison, W. R. (1995). Starch lipids and how they are related to starch granule structure and functionality. *Cereal Foods World,* **40**, 437-446.

Morrison, W.R. and Laignelet, B. (1983). An improved colorimetric procedure for determining apparent and total amylose content in cereal and other starches. *Journal of Cereal Science,* **1**, 9-20.

Muhr, A.H. and Blanshard, J.M.V. (1982). Effect of hydrostatic pressure on starch gelatinisation. *Carbohydrate Polymers,* **2**, 61-74.

Outtrup, H. (1986). Maltogenic amylase from *Bacillus stearothermophilus*. US Patent 4,604,355.

Outtrup, H. and Norman, B.E. (1984). Properties and application of a thermostable maltogenic amylase produced by a strain of *Bacillus* modified by recombinant-DNA techniques. *Starch,* **36**, 405-411.

Park, S.H., Wilson, J.D., Chung, O.K. and Seib, P.A. (2004). Size distribution and properties of wheat starch granules in relation to crumb grain score of pup-loaf bread (1). *Cereal Chemistry,* **81**, 699-704.

Pritchard, P.E. (1992). Studies on the bread-improving mechanism of fungal *alpha*-amylase. *Journal Biological Education,* **26**, 12-18.

Ring, S.G., Colonna, P., L'Anson, K.J., Kalichevesky, M.T., Miles, M.J., Morris, V.J. and Orford, P.D. (1987). Gelation and crystallisation of amylopectin. *Carbohydrate Polymers,* **162**, 277-293.

Russo, J.V. and Doe, C.A. (2005). Heat treatment of flour as an alternative to chlorination. *Journal of Food Technology,* **5**, 363-374.

References

Sahlstrom, S., Baevre, A.B. and Brathen, E. (2003a). Impact of starch properties on hearth bread characteristics. I. Starch in wheat flour. *Journal of Cereal Science*, **37**, 275-284.

Sahlstrom, S., Baevre, A.B. and Brathen, E. (2003b). Impact of starch properties on hearth bread characteristics. II. Purified A- and B-granule fractions. *Journal Cereal Science*, **37**, 285-293.

Sayre, R. N., Saunders, R. M., Enochian, R. V., Schultz, W. G. and Beagle, R. L. (1982). Review of rice bran stabilisation systems with emphasis on extrusion cooking. *Cereal Foods World*, **27**, 317-322.

Seetharaman, K., Yao, N. and Goff, E. T. (2002). Quality assurance for hard pretzel production. *Cereal Foods World*, **28** (5), 361-364.

Shiau, S.Y. and Yeh, A.I. (2001). Effects of alkali and acid on dough rheological properties and characteristics of extruded noodles. *Journal of Cereal Science*, **33**, 27-37.

Smith, A.M., Denyer, K. and Martin, C. (1997). The synthesis of the starch granule. *Annual Review of Plant Physiology & Plant Molecular Biology*, **48**, 67-87.

Snyder, E.M. (1984). Industrial microscopy of starches. In 'Starch Chemistry and Technology', R.L.Whistler, J.N. BeMiller and E.F.Paschall, Eds, Academic Press, New York, pp. 661-674.

Suggett, A., Ablett, S. and Lillford, P.J. (1976). Molecular motion and interactions in aqueous carbohydrate solutions. II. NMR studies. *Journal of Solution Chemistry*, **5**, 17-31.

Suggett, A. and Clark, A.H. (1976). Molecular motion and interactions in aqueous carbohydrate solutions. I. Dielectric-relaxation studies. *Journal of Solution Chemistry*, **5**, 1-15.

Takahashi, K., Matsuda, T. and Nitta, Y. (2001). Time-course changes of the fine structure of rice starch grains during cooking process by scanning electron microscopy. *Japanese Journal of Crop Science*, **70**, 47-53.

Tester, R.F. and Debon, S.J. (2000). Annealing of starch - a review. *Journal of Biological Macromolecules*, **27**, 1-12.

Tester, R.F., Karkalas, J. and Xin Qi (2004). Starch composition, fine structure and architecture. *Journal of Cereal Science*, **39**, 151-165.

Thomas, D.J. and Atwell, W.A. (1998). 'Starches'. AACC St. Paul, Minn, pp. 1-100.

Thomasson, C.A., Miller, R.A. and Hoseney, R.C. (1995). Replacement of chlorine treatment for cake flour. *Cereal Chemistry,* **72**, 616-620.

Van Bruijnsvoort, M., Wahlund, K.G., Nilsson, G. and Kok, W.T. (2001). Retention behaviour of amylopectins in asymmetrical flow field- flow fractionation studied by multi-angle light scattering detection. *Journal of Chromatography,* **A 925**, 171-182.

Vignaux, N, Doehlert, D.C, Elias, E.M., McMullen, M.S., Grant, L.A. and Kianian, S.F. (2005). Quality of spaghetti made from full and partial waxy durum wheat. *Cereal Chemistry,* **81** (1), 93-100.

Wasserman, L.A., Eiges, N.S., Koltysheva, G.I., Andreev, N.R., Karpov, V.G. and Yuryev,V.P. (2001). The application of different physical approaches for the description of structural features in wheat and rye starches. A DSC study. *Starch,* **53**, 629-634.

Weegels, P.L., Marseille, J.P, and Hamer, R.J. (1988). Gluten washing. *Staerke*, 40, 342-346.

Wiggins, C. (1998). Proving, baking and cooling. In The Technology of Breadmaking, S. Cauvain and L.S. Young, Eds. (London: Blackie Academic & Professional), pp. 120-148.

Williams, M.L., Landel, R.F. and Ferry, J.D. (1955). Temperature dependence of relaxation mechanisms in amorphous polymers and other glass forming liquids. *Journal of the American Chemical Society*, **77**, 3701-3706.

Wu, H.C.H. and Sarko, A. (1978a). The double helical molecular structure of crystalline α-amylose. *Carbohydrate Research* **61**, 7-26.

Wu, H.C.H. and Sarko, A. (1978b). The double helical molecular structure of crystalline β-amylose. *Carbohydrate Research* **61**, 27-40.

Zweifel, C., Handschin, S., Escher, F. and Conde-Petit, B. (2003). Influence of high temperature drying on structural and textural properties of durum wheat pasta. *Cereal Chemistry,* **80** (2), 159-167.

Chapter 9

INDEX

Acetylated distarch adipate	54
Acetylated distarch glycerol	54
Acetylated distarch phosphate	54
Acetylated starch	54
Acid thinning	48, 49
Amaranth	34, 36
Amylases; bacterial	50, 85
Amylases; cereal	85, 99, 110, 112
Amylases; fungal	110, 112
Amylases; maltogenic	112
Amylomaltase	51
Amylopectin	2, 3, 4, 6, 9, 12, 43, 60, 110, 112, 132, 137
Amylose	2, 3, 8, 17, 26, 60, 61, 102, 112, 132, 137
Amylose lipid complexes	112
Arrowroot flour	36
Barley; cooked pearled	36
Biscuit; cookie	120
Biscuit; cracker	122
Biscuit; semi-sweet	122
Biscuit; short dough	120
Biscuit; wafer	123
Bleaching	49, 50
Bread; croutons	130, 131
Bread; crust	111
Bread; naan	114
Bread; pita	114
Bread; rye	114
Bread; staling	111
Bread; standard UK	108
Bread; structure formation	108

Breadings; conventional	128
Breadings; cracker technology	129
Breadings; extruded	129
Breadings; Japanese Panko	128
Breadsticks; Grisini	130
Breakfast cereal; extruded	126
Breakfast cereal; flaked	124
Breakfast cereal; puffed grain	125
Breakfast cereal; hot water	125
Breakfast cereal; pellet process	127
Breakfast cereal; puffed pellet	125
British gums	49
Buckwheat flour,	36
Buckwheat grain	36
Cakes; high ratio	105
Cakes; low ratio	105
Cakes; recipe rules	104
Cakes; role of egg	107
Cakes; staling	107
Cold fruit pie filling	82
Cold processing	90
Condensation reactions	49
Confectionery	137
Cook-up starch	84
Critical concentration	85
Crystalline patterns	6
Custard; UHT	91
Custards; cook up	88
Custards; instant	90
Cyclodextrins	51
Dextrinisation	49
Distarch phosphate	52
Dry mixes	88
Enzyme hydrolysis	50
Epichlorohydrin	52
Extrusion cooking	121, 126, 127, 129, 132, 133,

Index

	134
Flan fillings	94
Fruit pie fillings	94
Glucoamylase	50
Glucose syrups	50
Gluten-free bread	113
Glycotransferases	51
Gravy powder	88
Growth rings	4
Heat treatment; cake flours	43
Heat treatment; drying	40
Heat treatment; pregels	41, 56, 57
Heat treatment; reducing microbial load	42
Heat treatment; steaming flour	42, 43
Heat treatment; steaming grain	42
Heat treatment; VHT starch	46
Hot roll cooking	57
Hydroxypropyl distarch adipate	54, 55
Hydroxypropyl distarch glycerol	54, 55
Hydroxypropyl distarch phosphate	54, 55
Hydroxypropyl starch	54, 55
Instant desserts	82
Instant starch	57, 90
Liquorice	139
Maize (*Zea Mays*)	36
Maize flour, masa	36
Maize grain	36
Maize grits, dry	36
Maize starch	36
Marshmallow	138
Measurement; amylopectin	61
Measurement; amylose	60
Measurement; composition	60
Measurement; crystallinity	65, 66
Measurement; gelatinisation temperature	65
Measurement; gel strength	75

Measurement; granule size	62
Measurement; viscosity	68
Meat paste	116
Milling; barley	22, 23
Milling; maize	21
Milling; oats	24
Milling; rice	23
Milling; rye	22
Milling; wheat	20
Naan bread	114
Noodles; alkaline	103
Noodles; plain wheat	103
Oat groats (*Avena sativa* L)	36
Oats, rolled or oatmeal,	36
Octenylsuccinate	56
Oxidation	49
Pan cooking	91
Pasta; cooked pasta	99
Pasta; dried	99
Pasta; fresh	99
Pasta; partial cooking	99
Pasta; traditional	98
Pasta; unconventional materials	101
Phosphated distarch phosphate	54
Pita bread	114
Popodoms	136
Potato: fried	97
Potato starch	30
Potato: boiled	96
Potato; flakes	41
Potato; granules	41
Pregelatinised	56
Pretzels	136
Propylene oxide	55
Proteins	11
Pullulanase	50
Pyrodextrins	49

Quinoa	36
Retorting	91
Retrogradation; high moisture	87
Retrogradation; intermediate moisture	107, 111
Retrogradation; low moisture	15, 137
Rice (*Oryza sativa* L)	36
Rice bran, crude	36
Rice flour, brown	36
Rice flour, white	36
Rice, brown, long grain,	36
Rice, white, glutinous	36
Rice, white, long grain,	36
Rice, white, medium grain	36
Rice, white, short grain	36
RVA; Final viscosity	71
RVA; Peak viscosity	71
RVA; Setback	71
Rye (*Secale cereale*)	36
Rye flour, dark	36
Rye flour, light	36
Rye flour, medium	36
Samosas	115
Sauce; liquid cooking sauce	81, 84
Sauce; prawn cocktail	90
Sauce; quick cook	88
Sauce; ready meal	91
Sauce; red and brown condiments	91
Sauce; UHT	91
Sauces; fruit, caramel, chocolate	91
Sausages	129
Snacks; crisps	131
Snacks; extruded direct	132
Snacks; extruded pellet	133
Snacks; fried dough	134
Snacks; popodoms	136
Snacks; pretzels	136
Snacks; puffed cereals	135

Snacks; rice cakes	135
Sodium trimetaphosphate	52, 56
Sorghum	36
Soup; canned	86, 91
Soup; cook up	89, 98
Soup; quick cook	88
Spray-drying	57
Spring rolls; rice based	116
Spring rolls; wheat based	115
Starch granules; formation	3
Starch granules; gelatinisation with sugars	13
Starch granules; high pressure gelatinisation	14
Starch granules; lamellar structure	7
Starch granules; non-starch components	11
Starch granules; normal gelatinisation	11
Starch granules; sizes and shapes	5, 34, 35
Starch granules; structure	4
Starch granules; swelling power and solubility	64
Starch manufacture; arrowroot	33
Starch manufacture; cassava	31
Starch manufacture; maize	26
Starch manufacture; potato	30
Starch manufacture; rice	29
Starch manufacture; sago	33
Starch manufacture; tapioca	31
Starch manufacture; wheat	28
Starch polymer crystalline structures	6
Starch polymers; glass transition	11
Starch synthetases	3, 9, 51
Succinylation	56
Tapioca, pearl, dry	36
Triticale, whole; grain flour	36
Vinyl acetate	55
Waxy maize	10, 39, 46, 53
Waxy potato	53
Waxy rice	18, 34
Waxy wheat	4, 98

Index

Wheat bran, crude	37
Wheat bulgur, dry	37
Wheat couscous, dry	37
Wheat flour all purpose	37
Wheat flour bread	37
Yogurt fruit fillings	2, 84, 91